Urban Greenway System Optimum Design

城市绿道系统优化设计

交通与发展政策研究所（中国办公室） 编著
Institute for Transportation &
Development Policy(ITDP-China)

江苏凤凰科学技术出版社

序

我非常荣幸向大家介绍《城市绿道系统优化设计》这本书。中国绿道建设始于广东，2010年开始在全国范围内得以迅速推广。兴建绿道的城市从最初广东省珠三角地区9个城市到现在北京、河北、福建等地区的六十多个城市。绿道的建设倡导了"绿色出行，低碳生活"的新生活理念，不但极大地改善了居民的生活质量，还给市民提供了舒适便捷的出行方式。尽管绿道发展在国内取得了巨大的成绩，但在"连通、连续、隔离、可达、服务"等方面的设计却普遍存在人性化设计缺失的问题，完善以上五个方面设计是保证绿道设计高品质的关键因素。

交通与发展政策研究所（Institute for Transportation and Development Policy，简称ITDP）成立于1985年，总部位于纽约，是一家非营利机构，其宗旨是推广在环境、经济、社会方面可持续发展的交通政策，并支持示范性交通运输方案，致力于公共交通（尤其是大运量公交系统）和非机动交通工具的推广、交通需求管理，以及土地利用规划的改善等。

ITDP近二十年来一直致力于绿道和慢行系统的研究及相关项目设计，拥有世界顶级的绿道专家团队。绿道在中国发展伊始，ITDP第一时间就介入到其规划、设计和建设进程中。在过去的几年里，ITDP一直致力于高品质绿道的设计，基于广州、东莞、惠州、天津、北京、武汉、宜昌等城市的实际情况，结合ITDP团队的国际绿道规划、

I am very pleased to introduce the book, *Urban Greenway System Optimum Design*. The construction of greenways in China originated from Guangdong and expanded rapidly in China since 2010. The first parts of cities with greenways constructed in China were the nine cities in the Pearl River Delta, Guangdong Province. More greenways were later built in Beijing and more than 60 cities in other provinces including Hebei and Fujian. Greenway construction, when done well, has the potential to not only greatly improve residents' quality of life, but also to put into practice the new concept of "green travel, low carbon life", providing a safe and convenient way to travel. Despite the rapid expansion of greenway construction in China, however, there are many common design problems relating to connection, continuity, separation accessibility and services. These are key areas of greenway performance, and improvements are needed as key factors to ensure high quality greenway designs.

The Institute for Transportation and Development Policy (ITDP) was founded in 1985 and had been headquartered in New York. IDTP is a non-profit organization whose purpose is to promote environmental, economic and social aspects of sustainable transport policies and support exemplary transport programs. ITDP dedicates itself to public transport (particularly mass transit systems), the promotion of non-motorized transport, transportation demand management, and improvement of land-use planning.

ITDP has carried out greenways and NMT research and project design for 20 years. Now ITDP has a world-class greenway expert team. As soon as the construction of greenways began in China, ITDP was involved in the design, planning and construction process. In the past few years, ITDP has been committed to enhancing good greenway design and planning in Chinese cities including Guangzhou, Dongguan, Huizhou, Tianjin, Beijing, Wuhan and Yichang drawing from local and international experts. This research includes development of urban greenway planning and design principles in several cities. ITDP hopes that these efforts will help in the process of greenway construction in China in the future.

ITDP's work is not only about the construction of greenways, but is also committed to

Preface

设计及建设经验,提出了适合中国城市发展的绿道规划与设计准则,希望对未来中国绿道的建设有所助益。

ITDP 的工作不仅限于绿道的建设,也致力于推动可持续交通在全球的发展。ITDP 在世界银行、欧洲复兴开发银行、亚洲开发银行定位城市交通政策时扮演着重要的角色,在联合国可持续发展委员会制定交通部分策略方面也发挥了重要的作用,并编写了联合国发展计划署的交通政策。2010 年 3 月,ITDP 在广州合法注册了广州市现代快速公交和可持续交通研究所,推动中国城市可持续发展的交通模式发展。目前 ITDP 已经与广州、兰州、天津、武汉、哈尔滨、宜昌等城市签订了合作备忘录,为他们的可持续交通项目(BRT、NMT、绿道、公共自行车和停车管理等)提供技术支持,其中广州及宜昌 BRT 交通走廊结合绿道的规划、设计及建设分别荣获多项国际大奖,也期待着未来更多的城市能够取得类似的成功。

卡尔·费
ITDP 东亚及东南亚区域负责人

promoting the development of sustainable transport around the world. ITDP plays an important role when the World Bank, the European Bank for Reconstruction and the Asian Development Bank set their urban transport policy. In addition, at the United Nations Commission on Sustainable Development, ITDP plays an important role in the strategy, and edits the transport policies of the United Nations Development Programme. In March 2010 ITDP was registered in China as the Guangzhou Modern BRT and Sustainable Transport Institute, in order to promote sustainable modes of transport and urban development. ITDP has signed cooperation memoranda with the cities of Guangzhou, Lanzhou, Tianjin, Wuhan, Harbin and Yichang, so that ITDP can provide technical support while these cities continue their traffic projects in areas of BRT, NMT, greenways, bike sharing and parking management. The Guangzhou and Yichang BRT corridors planned with key inputs from ITDP, which both include greenway components, have received major international prizes. ITDP looks forward to having the chance to achieve similar successes in more cities in the future.

Karl Fjellstrom
Regional Director, East & Southeast Asia, ITDP

CONTENTS 目录

第一章 概论
Chapter One Introduction

绿道的发展史 ········· 010
Development History of Greenway

第二章 现象与设计方法
Chapter Two Phenomenon and Design Method

1 中国绿道设计时五个最重要的技术问题 ········· 020
Five Most Important Technical Questions

1.1 绿道的连通 ········· 020
The Connectivity of Greenway

1.2 绿道的连续 ········· 023
The Continuity of Greenway

1.3 绿道的隔离 ········· 032
The Separation of Greenway

1.4 绿道的可达 ········· 038
The Accessibility of Greenway

1.5 绿道的服务 ········· 042
The Service of Greenway

2 绿道设计方法 ········· 047
The Design Method of Greenway

2.1 网络规划 ········· 047
Network Planning

2.2 基础设施设计 ········· 052
Infrastructure Design

2.3 配套设施设置 ········· 069
Supporting Facility Setting

3 小结 ········· 075
Conclusion

第三章 实践案例
Chapter Three Practice

荔湾旧城慢行系统改善建议
Liwan Old Town NMT Improvement Suggestion

1 完善高质量网络 ······ 078
To Optimize High Quality Network

1.1 人行道 ······ 078
Sidewalk

1.2 自行车通行环境 ······ 085
Cycling

1.3 交叉口 ······ 096
Intersection

1.4 路中过街 ······ 106
Mid-Block Crossing

1.5 行人导向系统 ······ 114
Pedestrian Guidance System

2 综合改造方案 ······ 120
Comprehensive Proposals

珠江新城绿道改善建议
Zhujiang New Town Greenway Improvement Suggestion

1 概况 ······ 124
Overview

2 现状问题分析 ······ 126
Current Condition Analysis

2.1 现状概述 ······ 126
Current Condition Overview

2.2 现状典型问题及改善建议 ······ 128
Current Issues and Improvement Suggestions

3 总体规划布局 ······ 134
Overall Layout

3.1 完善绿道网络 ······ 134
Improve Greenway Network

3.2 行人轨迹调查 ······ 138
Pedestrian Trajectory Survey

3.3 构建密集的街道网络 ······ 142
Build Dense Street Network

4 道路横断面改善建议 ---------- **144**
Road Cross Section Improvement Suggestion

4.1 道路断面现状 ---------- 144
Current Condition of Road Section

4.2 道路断面改善建议 ---------- 146
Road Section Improvement Suggestion

5 过街通道改善建议 ---------- **150**
Street Crossing Improvement Suggestion

5.1 过街通道现状分析 ---------- 150
Analysis of the Current Street Crossing Situation

5.2 平面过街和立体过街 ---------- 153
Crosswalk and Pedestrian Bridge or Tunnel

5.3 交叉口改善建议 ---------- 156
Intersection Improvement Suggestions

6 配套设施建设 ---------- **158**
Facilities Construction

6.1 自行车停车点 ---------- 158
Bicycle Parking Point

6.2 公共自行车 ---------- 163
Public Bicycle

兰州绿道改善建议
Lanzhou Greenway Improvement Suggestion

参考方法 ---------- 166
Reference Method

广州淘金-建设新村改造建议
Taojin-Jianshe Xincun Reform Suggestion

1 项目区位 ---------- **178**
Project Location

2 道路网络分析及建议 ---------- **179**
Road Network Analysis and Suggestion

2.1 道路网结构 ---------- 179
Road Network Structure

2.2 道路流量分析 ---------- 180
Road Traffic Analysis

2.3 道路交通组织优化 ---------- 182
Road Traffic Organization Optimization

3 道路平面及横断面设计 ---------- **184**
Road Plan and Cross Section Design

3.1 设计要点 ---------- 184
Design Key Point

3.2 道路平面及横断面设计 ---------- 189
Road Plan and Cross Section Design

老挝万象绿道规划设计
Greenway Planning and Design in Vientiane, Laos

1 目标 ··· 214
Objectives

2 当前问题 ································· 214
Current Problems

2.1 缺失及不连续的人行道 ············ 214
Missing and Discontinuous Walkways

2.2 障碍物 ································· 216
Obstacles

2.3 人行道上的停车 ····················· 217
Parking on Sidewalks

2.4 交叉口 ································· 217
Intersections

2.5 人行横道 ······························ 218
Pedestrian Crossings

2.6 缺乏遮盖设施 ························ 218
Lack of Shade

3 建议书 ······································ 219
Proposals

3.1 人行道 ································· 219
Sidewalks

3.2 交叉口和过街通道 ·················· 222
Intersections and Crossings

3.3 道路设计和宁静交通 ··············· 223
Road Layouts and Traffic Calming

3.4 街道家具和照明设备及景观美化 ·· 224
Furniture and Light and Landscaping

3.5 公共座椅 ······························ 225
Public Seating

3.6 树木 ···································· 225
Trees

作者简介 ······································ 226
About the Authors

第一章
Chapter One

概论
Introduction

Chapter One

绿道的发展史
Development History of Green

世界上第一条真正意义的绿道始建于1867年，是Frederick Law Olmsted 设计的美国波士顿公园绿道。而1996年制定完成的《泛欧生态和景观多样性战略》，为欧洲各国协调绿道规划建设提供了基础性框架。经过一个多世纪的理论探索与建设实践，特别是20世纪80年代绿道得名以来，建设绿道成为一个国际运动，在全世界蓬勃发展，世界上有数千个国际、国家和区域层次的绿道项目。绿道系统的规划建设也逐渐成熟和完善，并成为世界各国解决生态环保问题和提高居民生活质量的重要手段。目前，主要发达国家基本上都进行了城市绿道系统建设。

The world's first greenway was constructed in 1867. It was the Boston Public Garden greenway in the United States, designed by Frederick Law Olmsted. *The Pan-European Biological and Landscape Diversity Strategy*, made in 1996 provided a fundamental framework for the greenway construction in various European countries. After one century's theoretical research and constructive practice, especially after the definition of greenway in 1980s, greenway has become an international movement and is developed in fast speed. The world has thousands international, national and regional greenways. The construction of greenway system gradually become mature and improved. It has become a significant measure to enhance civilian's living standard and to solve the ecological and environmental issues.

中国的绿道建设起步较晚。2010年，广东省批准《珠江三角洲绿道网总体规划纲要》，在此规划里国内第一次出现了绿道的定义，开始了中国绿道的建设步伐。广东省最早的绿道建设主要为景观类绿道的建设，大部分绿道设置在实施条件较好的城市外围地区。这些绿道在投入使用后出现了一个难以回避的问题：尽管拥有高质量的绿道，但是由于绿道距离社区过远，居民到达不便，导致绿道使用率偏低。而在市区内部距离社区较近的绿道却很受市民欢迎。

The greenway construction in China was carried out in a late start. In China, the construction of greenway started after *The Pearl River Delta Greenway Network Master Plan Outline*, which was sanctioned by Guangdong province in 2010. The beginning of greenway construction in Guangdong

province was mainly constructions inside landscapes. Most of the greenways were implemented in the outer areas of cities where is easier for the construction. There is an inevitable task following the openning of the greenways. Despite the high quality of greenways, they are not convenient for civilians to access, thus causes low frequency of usage. However, the greenways located inside the cities near the neighborhoods are very popular among the civilians.

2010年后，越来越多的城市开始意识到绿道对城市发展有着巨大的正面影响，绿道也逐渐成为城市建设的宠儿，在中国快速发展起来。截止到2014年底，中国绿道建设总里程已达到10950公里，规划建设总里程5万余公里，兴建绿道的城市从最初广东省珠三角9个城市到现在广东、北京、河北、福建等数省的六十几个城市。尽管国内绿道、省内绿道建设取得了非凡成绩，但是与国外优秀绿道系统相比，普遍在"连通、连续、隔离、可达、服务"等几个人性化设计方面存在问题，这五个方面是一个良好绿道系统的关键因素，也是未来中国绿道需要着重完善和努力的方向。

After 2010, more and more cities started to realize that greenway has great positive influence on cities' development. Greenway has gradually become the favor of city construction, and has been developing rapidly in China. By the end of 2014, the total greenway construction has reached almost 10,950 km. The expected total length is 50,000 km. The cities have newly constructed greenway from the earliest 9 cities in Pearl River Delta in Guangdong to Beijing, Hebei, Fujian and over 60 cities in other provinces. Although the provincial greenway construction in domestic greenway has gained remarkable results, however, there are still some problems compared with the greenway system overseas, such as the issues of connectivity, continuity, accessibility, services and other problems about human-oriented design. These are key elements for constructing a good greenway system, and are also the important points that should be enhanced in the future greenway in China.

Chapter One

国内外目前在绿道理论研究方面，涌现出了大量的研究成果，出版了大量的研究专著，但是绿道关于使用者的需求和人性化的设计方面的绿道文献却并不多见，因而具体的技术细节也并不为国内多数人知晓。本书所介绍的设计方法，主要基于ITDP的国内外绿道规划建设经验和对国外绿道设计细节的研究，这些方法未必覆盖绿道设计的所有方法，但却是国内较为少见的方式，既简单易行、成本低，又针对国内现状，效果显著。对于提高绿道安全性，建设便捷和易于使用的绿道，有着重要的现实意义。希望这些设计手法能够广泛应用于中国的绿道建设之中，建设更有吸引力的中国绿道。

Currently, there are a large number of research achievement and publishment on theoretical analysis of greenway in China and other countries. However, research on users' needs and human-oriented designs in greenway are not sufficient, thus, specific technologies and detail designs are not widely known in China. This book introduces the design measures, which are mainly based on ITDP's greenway planning and constructing experience domestically and internationally, and researches on international detail designing. These measures do not fully cover all design methods of greenway, but those are relatively rare in China. They are easy for implementation and cost-saving. They also fit the current situation in China and have effective results. Therefore, they have realistic significance on enhancing the safety, convenience and accessibility of greenway. We hope that these design measures can be widely utilized on the greenway construction in China, and help our country to make greenway more attractive.

概论
Introduction

什么是绿道

What is Greenway

如果让十个不同的人来描述绿道，可能会有十个不同的答案。这是因为，绿道是一个通用的概念，绿道可以是任何一种我们认可的样子。而从绿道发展的历程上来看，正是由于绿道这一概念的开放性，在不同的时代、不同的地区，人们对于发展的目标、面临的资源限制、游憩娱乐的需要都不尽相同，所以在全球范围内建设了多种形式的绿道系统。经过多年的发展，目前在世界范围内，对于绿道的理解，有以下几种典型的类型。

People have different explanations on greenway. That is because greenway is a general concept, and it could be any concept which we depicted. From the history of the development of greenway, the concept of greenway is very wide. In different times and areas, people have different limited resources, different purposes of development and different demands for recreation. Therefore, various greenway systems are constructed globally. After many years of development, the comprehension on the concept of greenway has the following types in the world.

美国绿道

American Greenway

绿道（Greenway）是一种线形开敞空间，它通常沿着自然廊道建设，如河岸、河谷、山脉或者是在陆地上沿着由铁道改造而成的游憩娱乐通道、一条运河、一条景观道路或其他线路。绿道开放空间把公园、自然保护区、风景名胜区、历史古迹及人口密集地区等连接起来。

Greenway is a linear open space, which is usually constructed along the natural corridor, such as rivers, valleys, mountains, a recreational corridor transfered from railway, a canal, a landscaping road and other routes. The open space in greenway connects parks, natural reservation zones, scenic spots, historical sites and densely populated areas.

Chapter One

欧洲绿道

European Greenway

欧洲绿道联合会 EGWA 于 2000 年对绿道作了如下的界定：①专门用于轻型非机动车的运输线路；②已被开发成以游憩为目的，或为了承担必要的日常往返需要(上班、上学、购物等)的交通线路，一般提倡采用公共交通工具；③处于特殊位置的、部分或完全退役的、曾经被较好恢复的上述交通线路，被改造成适合于非机动交通的使用者使用的，比如徒步者、骑自行车者、限制性机动者（指被限速或特指类型的机动车）、轮滑者、滑雪者、骑马者等。

European Greenway Association (EGWA) has developed the following definitions in 2000. ① Specifically for light non-motorized traffics. ② It has been developed into a recreational purposes and/or undertaken the necessarily daily commute (to work, school and shopping, etc.), which generally advocates the use of public transport. ③ The transportation lines locate in special areas, or have completely retired and was transformed into a suitable transportation routes, which can be reconstructed for the users of non-motorized transportation, including hikers, cyclists, restricted vehicles (limited speed or other specific types of motor vehicles), roller skating, skiers and horse riders.

中国绿道

China Greenway

在中国,绿道的发展始于2010年的广东省绿道建设,国内对绿道的定义基本上源于广东省的定义。《广东省省立绿道建设指引》中,绿道的概念:"绿道是一种线形绿色开敞空间,通常沿着河滨、溪谷、山脊、风景道路等自然和人工廊道建立,内设可供行人和骑车者进入的景观游憩线路,连接主要的公园、自然保护区、风景名胜区、历史古迹和城乡居住区等。"这个定义实际上是很大程度上借鉴美国绿道的定义。

In China, the development of greenway started in the construction of greenway in Guangdong Province. The definition is basically originated from Guangdong. In *Benchmark Technical Regulations of Regional Greenways* in Guangdong Province, the concept of greenway is defined: Greenway is a green linear open space, generally constructed along the rivers, valleys, mountains, landscapes and other artificial corridors. Within the greenway, landscape lines for pedestrian and cyclist can be provided to connect parks, natural reserves zone, famous scenic spots, historical monuments and urban and rural residential areas, etc.. This definition is, in fact, largely drawn from the concept of the American greenway.

在中国,绿道的定义也是随着实际需求的变化而改变的,这也是绿道在中国发展中呈现出的一个特点。在总结绿道建设的经验后,中国也逐渐赋予了中国绿道新的内涵。根据珠三角前期绿道建设经验,2011年广州市绿道建设提出三级网络理念:即省绿道、城市绿道和社区绿道,并将绿道延伸至社区内部,在营造良好景观环境的同时,更好地为居民提供出行便利。这就意味着绿道系统开始承担起改善居民出行环境、支持步行和自行车交通的功能,这些经验也为后来其他城市的绿道建设提供了借鉴。绿道建设开始更多地考虑使用者的便捷性,越来越多的城市在市区内部设置城市绿道。随着绿道系统的投入使用,绿道在城市中的功能也从原来的单纯景观功能向交通功能转变。而这一经验,也被中国其他地区在建设绿道时迅速借鉴。

In China, the definition of greenway changes according to actual needs, which is a characteristic presented during the development. After summing up the constructing experience, Chinese government gradually endows greenways with new connotation. According to the greenway constructing experiences in earlier stage of Pearl River Delta, Guangzhou greenway construction put forward the idea of three-hierarchy network in 2011: provincial greenways, urban greenways and community greenways. These three types of greenways are all extended to the internal of communities to build better landscape at the same time provide a better convenience for residents' travel. This means greenway system began to take the functions of improving traveling condition and supporting pedestrian or cycling transportation, and these experiences were learned by other cities afterward. Convenience is getting more important in greenway construction while more and more urban greenways are built inside the downtown. With the greenway system coming into service, the functions of greenways in cities changes from simple view function to traffic function. This experience is also learned by other areas in China.

Chapter One

ITDP 中国绿道定义

ITDP Chinese Greenway Definition

正是由于绿道是一个不断变化的概念，单纯对绿道进行定义确实有局限性。绿道的实质是通过提供多种联系促成多类别的活动，满足具体的、不同种类使用者需求。结合中国目前绿道建设来看，对于中国这样的大国，地区差异比较明显，在规划建设绿道时，应不拘泥于绿道的形式，而注重绿道的实际功能和用途，能够弹性的适应政府的需要和公众的需求。

It is because the definition is constantly changing, the simple definition for greenway has limited explanation. The essence of greenway is to provide various connections for many types of activities, and satisfy different groups of users' requests. As for the current greenway construction in China, the greenway construction is determined by the various differences in many regions in this big country. The greenway construction should not adhere to the form of greenway, but pay attention to the practical function, and adapt to the needs and demands of governments and the public.

因此，ITDP 认为：高质量的非机动车交通廊道，通过人性化的设计，让使用者能便捷、安全使用，即为绿道。

Therefore, ITDP believes that the greenway can be defined as "high quality non-motorized traffic corridor that provides convenience and safety to users through human-oriented design."

概论
Introduction

第二章
Chapter Two

现象与设计方法
Phenomenon and Design Method

Chapter Two

1. 中国绿道设计时五个最重要的技术问题
Five Most Important Technical Questions

绿道在中国已经得到广泛的应用，在短短几年的时间内，取得了巨大的成功。但是从使用者的体验感和提高绿道吸引力的角度来看，如果在绿道规划设计时很好的解决以下五个方面的问题，将会让绿道实现更大的社会效益，这五个方面的问题是"连通、连续、隔离、可达、服务"。

Greenway has been widely implemented in China and achieved a great success in just a few years. For the consideration of users' experiences and improving the attractiveness of greenway, the following five aspects which are 'connectivity, continuity, separation, accessibility and service' need to be properly accommodated during planning and designing to achieve greater social effects.

1.1 绿道的连通
The Connectivity of Greenway

使用绿道的方式主要是步行或者骑行，步行和骑行是体力活动，因此对距离的敏感度非常高，出行意愿和舒适性容易受到绕行路径的影响。因而在绿道网络规划时，必须保证绿道的连通性，绿道网络应该贴近需求，直接通往所有的目的地，连接重要的节点和人流集中的地点，并且线路的选择还可以更加的有趣，增加人们的体验感。同时，绿道系统必须成网，相互之间有效的连接成为一个有机的整体。

People use greenway most for walking or cycling which are both physical activities with a very high sensitivity with distance, so the travel willingness and comfort are easily influenced by circuitous roads. Therefore, its connectivity must be ensured when greenway network is planned at the same

广州荔枝湾涌绿道，连接了周边的社区，吸引了大量的市民使用

Lizhi Wan Chong greenway in Guangzhou connects communities around and attracts lots of citizens to come.

time close to the need and lead directly to all destinations connecting important joints and sites of high population, on the other hand the choice of routes may be more interesting to enhance people's experience. Meanwhile, all the greenways need to be planned to form a net where greenways link each other effectively to become an organic whole.

绿道网络规划也应符合实际的使用需求，在需求较高的区域，绿道网络密度应规划得更为密集，而在需求不高的区域，满足绿道网络的有效连通即可。广东省最开始的绿道建设主要为景观内绿道的建设，部分绿道设置在实施条件较好的城市外围地区，这些可达性低的绿道，使用率也大大降低，一定程度上浪费了城市基础建设投入，又无法提高人们的生活品质。在国内，不难发现一些建设标准很高的绿道，鲜有使用者，形同装饰品，大多是因为这些绿道既不连接重要的目的地，周边也缺乏实际的需求。

Greenway network should be planned to meet practical use demands. In areas of high demand greenway is advised to be planned for a denser network, while in areas of low demand it is acceptable for greenway to just satisfy effective connection of network. At the very beginning greenways in Guangdong were mainly built in landscape with some of which were planned at the outskirts with better implementation condition. These greenways with poor accessibility waste the investment of urban infrastructure construction at a certain degree at the same time fail to improve people,s life quality because of their low service efficiency. In China, it is not difficult to find greenways with high construction standard but few people use them, making them no more than some embellishments. This is because that these greenways neither connects important destinations nor practical requirements around are needed.

广州萝岗生物岛绿道，尽管拥有高质量的绿道，但是由于相对独立，居民到达不便，导致绿道使用率偏低。

Luogang Bio-island greenway in Guangzhou. Although the greenway is of high quality, its relatively isolation and inconvenience for residents to access lead to a low usage rate.

纽约市绿道网络，在需求集中的城市中心区，规划设计了密集的绿道网络，而在城市的外围区域，绿道网络密度明显降低。

Greenway network in NY. Dense greenway network is planned in urban center where demand is intensive while the network density obviously reduces at the outskirts areas of the city.

Chapter Two

河流、铁路及高速公路等地理要素和基础设施常常成为绿道连通的障碍，这就需要设置桥梁来解决问题。国内许多城市的绿道系统因规划实施的协调不足或因江河、高等级道路的隔离而无法形成完整的网络。国内设置桥梁的时候，很少考虑设置专用的自行车和行人跨河桥梁。机动车桥梁上通常没有设置专用的自行车道，由于江河有船只通航的需求，许多桥梁净空较高，因此设置了长而蜿蜒的匝道，自行车难以通行；而人行天桥的坡道又过于陡峭，自行车只能推行上桥。因此，行人，甚至是部分自行车，宁愿冒险横跨护栏过马路，也不愿意使用天桥。

Geographic elements and infrastructures like rivers, railways and highways are often obstacles of greenway connection and this is where we need to set up bridges to solve the problems. Greenway systems in many cities of China cannot form complete networks since planning implementation is not coordinating or networks are blocked by rivers and roads of upper level. The setting of specialized bicycle and pedestrian crossing bridges is rarely taken into consideration when planning bridges. Rather than setting specialized bikeway, many motor vehicle bridges with high clearance for shipping navigation are planned long and winding ramps which are difficult for bicycles to go through, on the other hand ramps of pedestrian over crossing are too steep for bicycle that people have to get off their bikes and push to go through the bridge. Therefore, pedestrians and even parts of bicycle take risk to cross the road by stretching over the barrier rather than use the overbridge.

伦敦的千禧桥为步行、自行车专用的绿桥，连接泰晤士河南北两岸。

London Millennium Bridge is a green bridge for pedestrians and cyclists, connecting south and north banks of the Thanmes.

广州东濠涌绿道北段，两侧绿道被城市主干道隔断，连接不便，即使是外籍绿道使用者，也会选择横穿马路走到另外一侧绿道。

North section of Donghao Chong greenway in Guangzhou. Both sides of greenway are cut off by urban main road. With its inconvenient connection, even foreigners choose to cross the street directly to get to the other side of greenway.

乌得勒支某绿道连接桥梁，同时服务机动车、自行车及行人，但是桥梁的设计充分满足了自行车的出行需求，道路坡度也很平缓，骑行并不困难。

In Utrecht, the bridge connecting greenway serves motor vehicles, bicycles and pedestrians at the same time, but its design fully satisfies the travel demand of bicycles with gentle road grade which is easy for riding.

现象与设计方法
Phenomenon and Design Method

1.2 绿道的连续
The Continuity of Greenway

绿道网络的不连续，会严重影响绿道功能的发挥，即使我们完成了 99% 的绿道建设，其功能还是可能会因为 1% 的中断大打折扣。绿道空间被机动车违规占用，或被市政设备（电箱、灯杆、市政检查井等）和占道摆卖等侵占，使步行和骑行处处受阻。行人在横过马路的时候缺乏保护、自行车没有专用的过街通道、机动车出入口的影响，成为影响绿道连续的最常见的情况。

The discontinuity of greenway network seriously impacts its performance which declines greatly for its 1% of interruption despite of 99% of finished construction. The greenway space is occupied illegally by motor vehicles or invaded by municipal facilities (electric box, lamp pole, inspection well, etc) or street vendors to restrict walking and cycling. The most common situation to pull down the continuation of greenway is that people lack of protection and bicycles lack of special passageway when crossing the street with the vehicle access blocking the greenway.

中山大道绿道，实际调查中显示绿道的自行车道在多处并不连续。

The bikeway in Zhongshan Avenue greenway is discontinuous in many places according to actual investigation.

023

Chapter Two

骑自行车的人不愿意使用不合标准的、维修不善的、有障碍物的、狭窄的自行车道。在进行绿道设计时，必须要确保高质量的自行车道，必要时需要重新设计道路。

People who ride bicycles will not use narrow, sub-standard bikeway in poor maintenance with barriers on it. When planning greenway, one must ensure high quality bikeway and even redesign the whole route if necessary.

广州某绿道被机动车停车和垃圾转运车阻隔。
This greenway in Guangzhou is obstructed by illegal parking and garbage truck.

广州某绿道被临时建筑围蔽占用。
This greenway in Guangzhou is occupied by temporary building.

广州某绿道在人行天桥处被隔断，自行车被迫与机动车混行。
This greenway in Guangzhou is cut off at passenger foot-bridge causing the situation that bicycles are forced to mix with motor vehicles.

广州某绿道被标志牌完全隔断。
This greenway in Guangzhou is completely cut off by signboards.

现象与设计方法
Phenomenon and Design Method

广州某绿道被机动车停车和垃圾桶阻隔。

This greenway in Guangzhou is obstructed by parking and garbage cans.

兰州某绿道自行车道路面维修,自行车道无法使用。

The bikeway of greenway is not available as it is under maintenance in Lanzhou.

兰州某绿道自行车道因沿街摊贩占道和地下通道出入口中断。

This greenway in Lanzhou is interrupted by street vendors and underpass entrance.

Chapter Two

交叉口是行人、自行车与机动车的冲突点，大约80%的事故发生在交叉口，交叉口的设计关系到行人与骑行者的安全，也是绿道设计中无法回避的关键节点。

Intersection is the conflict point between pedestrians, bicycles and motor vehicles that causes about 80% of traffic accidents. Intersection design not only relates to the safety of both pedestrians and cyclists, but is also the unavoidable key point during greenway design.

国内的道路交叉口现状，部分没有对交通空间进行严格的限定，使道路使用者在通过的时候或是无所适从、或是随意穿行，形成了大量的冲突点，危及使用者的安全。许多交叉口尺度巨大，行人斑马线横穿数条车道，行人过街距离长，有的甚至不能在一个信号灯周期内过街，或没有行人的信号灯相位，中间并没有安全岛庇护行人，使行人完全暴露在车流中。而且国内道路交叉口很少设有自行车专用过街通道，大部分自行车与行人共用斑马线、天桥及安全岛。

Some of road intersections in China do not strictly define the traffic space, making road users either lost at the intersection or pass through at will when passing across to form lots of conflict points which threaten users' safety. Zebra crossing of many intersections go over several lanes with giant dimension, making crossing distance too far for people to cross the street in even one signal lamp cycle, or exposing pedestrians completely to the traffic stream as it lack of signal lamp phase or safety island. Furthermore, seldom intersections in China are equipped with special bikeway so that most of bicycles share zebra crossing, passenger foot-bridge and safety island with pedestrians.

广州中山大道绿道交叉口，骑行者必须与行人混行过街。

Cyclists have to cross the street mixing with pedestrians at the intersection of Zhongshan Avenue Greenway

交叉口处蓝色的自行车道涂装增加了驾驶员的注意力、提升了自行车的安全。

The bicycle lane painted in blue at the intersection helps to increase the attention of drivers and to improve riding safety

现象与设计方法
Phenomenon and Design Method

东莞某绿道有独立的自行车过街通道，但是要求骑行者推车过交叉口。

This greenway in Dongguan is equipped with independent bikeway, but cyclists are required to get off the bike pushing them across the street.

广州某绿道自行车道在交叉口处被中断。

The bikeway of greenway in Guangzhou is interrupted at the intersection.

目前绿道沿线交叉口，基本上沿用原有道路交叉口，并未因为绿道而对交叉口进行改善性的设计，导致对行人和自行车而言，使用绿道过街有时并不安全。同时，很多城市要求骑行者通过交叉口的时候下车推行，这也降低了绿道骑行的体验感。应在绿道过街处针对性的设置一些安全设施，像自行车过节通道、安全岛、机动车减速设施等，保护使用者的安全，确保所有使用者，包括儿童、老人和残疾人，都可以很好、很安全的使用绿道。

Now greenways at intersections basically still use the original intersection without an improved design which is sometimes unsafe for pedestrians and cyclists to cross the street, meanwhile it also impacts riding experience in greenway as cyclists are asked to get off the bike pushing it across the intersection in many cities. Some safety facilities like refuge island and speed bumps should be set at greenway intersection to ensure users' safety including children, elderly and the disabled who can all use the greenway well and safely.

Chapter Two

绿道应保持其自身的连续性，特别是在复杂的交通条件下，更是需要保证其整体连续。绿道可根据不同的现状情况做出具有适应性的设计，如下图阿姆斯特丹的绿道，在绿道过街处，结合现有流量，设置独立的自行车道和人行过街通道，并设置路中安全岛、独立的行人信号灯和自行车信号灯，确保行人和自行车可以便捷和安全的过街。

Greenway should be continued. Especially under complicated traffic conditions, it is necessary to keep its continuity. Adaptable design method can be used based on different situations. Shown in the following picture, designed according to the traffic volume, the greenway in Amsterdam has crosswalk, dedicated bike crossing, refuge island, pedestrian signal and bike signal, which ensures convenient and safe crossing of pedestrians and cyclists.

阿姆斯特丹某绿道交叉口，交叉口过街人行道和自行车道的设置是分别设置，交叉口也设置有路中安全岛，也设有专用的自行车和行人信号灯。

The greenway intersection in Amsterdam. Sidewalk and bikeway are set separately at intersection with central refuge as well as special signal lamp for cyclists and pedestrians.

巴塞罗那某绿道过街的通道采用颜色鲜明的涂装，非常清晰。

Crossing way of greenway in Barcelona is painted with distinct painting.

现象与设计方法
Phenomenon and Design Method

在绿道通过大型的渠化交叉口时，应设置专用的自行车过街通道，让行人、自行车过街时各行其道。在过街距离较长的路段，应设置路中过街安全岛，保障行人和自行车在路中的过街安全。自行车道在交叉口的路径应用颜色涂装，引导自行车过街和警示车辆减速。必要时可调整道路路缘石半径，采用"小半径 + 大半径"的设计，降低机动车横穿通过自行车道和人行道的车速，或采用抬升车道的方式，降低机动车车速。

When going through large-scale canalized intersections, special bikeway needs to be set to have pedestrian motor vehicle prevails with safety island being planned in the long-distance crossing to ensure pedestrians and cyclists passing the street safely. Bikeway at the intersection should be painted with different color to guide cyclists crossing the street at the same time warn vehicles to slow down. The radius of extroverted kerb can be adjusted when necessary adopting the design of "sharp radius + smooth radius" to slow down the speed of vehicles when passing bikeway and sidewalk or to slow down speed by lifting ane.

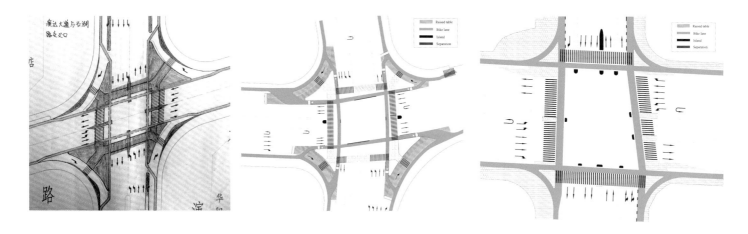

收窄交叉口，并设置环岛组织机动车交通流、自行车流和人流，在绿道设置专用自行车道，并调整通往绿道道路的设置，在交叉口路中设置路中安全岛，保护行人过街安全。

Narrowing the intersection and setting up roundabout to organize vehicle flow, bicycle flow and pedestrian flow with dedicated bikeway being planned in the greenway and to adjust the road setting toward the greenway. Ensure pedestrian crossing safety by setting up dedicated bikeway and central refuge island in the middle of intersection.

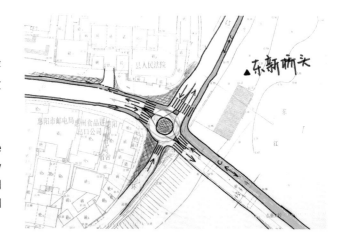

Chapter Two

外延交叉口人行道，采用"小半径 + 大半径"的设计，降低车辆车速；设置专用自行车过街通道，并在路中设置过街安全岛，保护绿道行人及自行车过街安全。增设路面的平面过街，利用现有交叉口中机动车无法使用的空间设置路中安全岛，保障绿道过街安全。

The extensional intersection sidewalk slows down the speed of right-turning vehicle with design of "sharp radius + smooth radius" at the same time setting up dedicated bikeway and central refuge island in the middle of intersection to ensure pedestrian crossing safety. To increase the construction of at-grade crossing, setting up central refuge island in the space where vehicles cannot reach of existing intersections to ensure crossing safety.

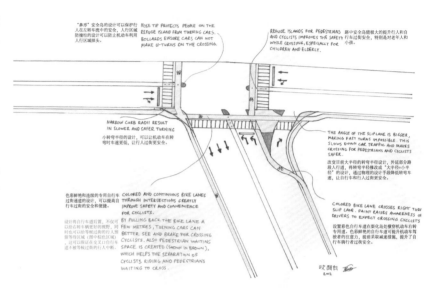

机动车出入口的设计，在国内几乎所有绿道建设中均未被引起足够的重视。骑自行车的人对于道路路况比乘坐汽车的人更为敏感，并且更喜欢平坦的路面。绿道应享有比机动车道更高的路权，否则机动车出入口频繁切断自行车道，即便设置有无障碍坡道，频繁的起伏也会严重影响骑行的舒适度。

Almost all the designs of vehicle access of greenway construction in China are not attached enough importance. Cyclists prefer to flat surface with higher sensitivity compared with vehicles drivers and passengers. Greenways should enjoy higher road right than motorways, otherwise frequent fluctuation that bikeway being cut off by vehicle accesses will seriously impact the riding comfort even though barrier-free ramps are planned.

广州中山大道绿道由于频繁被机动车出入口打断，绿道骑行的体验感明显降低。

Zhongshan Avenue greenway is frequently cut off by vehicle accesses leading to an obvious reduction of riding experience.

现象与设计方法
Phenomenon and Design Method

佛山某绿道自行车道由于高差，实际上无法连续。

The bikeway of greenway in Foshan is not continuous in fact as the altitude difference.

广州珠江新城某绿道被出入口中断。

Zhujiang New Town greenway is cut off by the access.

荷兰乌得勒支将机动车出入口提升，保持绿道的连续。

In Utrecht the Netherlands, the continuity of greenway is maintained by lifting up the vehicle access.

瑞典马尔默市为保持绿道在交叉口的连续，将某绿道在交叉口处整体提升。

In Malmo Sweden, the greenway at the intersection is wholly lifted to keep its continuity

Chapter Two

1.3 绿道的隔离
The Separation of Greenway

佛山某绿道仅是形式上的自行车道，无法真正意义上实现自行车道和人行道的隔离。

The greenway cannot achieve true separation between bikeway and sidewalk with only a formal bikeway in Foshan.

自行车和行人使用绿道时有着不同的体验方式和行为，步行者可能会向往美好的蜿蜒曲折的小径，而骑行者想要直接往前走，并不希望在树木和公园长椅周围绕来绕去。同时，从路径的材料来看，骑行者更喜欢平坦的路面，而步行者或许更喜欢美观的鹅卵石和粗砖路面。

Pedestrians and cyclists have different ways of experience and behaviors when using greenways, with the former might be look forward to beautifully winding path while the latter prefer to ride straight forward without circling around trees and benches. Meanwhile, from the perspective of path materials, cyclists prefer to flat surface while pedestrians are more likely to enjoy nice pebbles and rough bricks.

绿道的隔离的关键是处理机动车、自行车与行人之间的冲突，其最重要的措施是实施明确的功能分区，保证三者各行其道，从而减少冲突。另外，降低机动车的车速，也有助于降低非机动车交通的风险。

The key of greenway separation is to deal with the conflict among vehicles, bicycles and pedestrians with the most important measure which implements specific function division to ensure all three go their own ways to reduce conflict. Besides, slowing down vehicle speed contributes to lower the risk of non-motorized traffic.

广州某绿道在无物理隔离的情况下，自行车道的专用将很难得到保障。

The specialization of bikeway of greenway is hard to be ensured without physical separation in Guangzhou.

在绿道设计时，首先确保机动车道与绿道隔离，保障使用绿道的行人及自行车的安全，其次根据现状条件，隔离自行车道和人行道，如果有条件，设置分隔带分离自行车道和人行道，如果没有条件，应把自行车道和人行道在平面上分离。

When designing greenways, the first thing is to insure the separation between motorways and greenways to ensure the safety of pedestrians and cyclists who use the greenway and the second thing is to segregate bikeway and sidewalks according to existing conditions. To segregate bikeway and sidewalks if possible or to have them separated in the plane at least.

现象与设计方法
Phenomenon and Design Method

广州萝岗某绿道画线自行车道完全被机动车停车侵占。

The specified bikeway in Luogang is completely occupied by motors.

武汉某绿道画线自行车道被机动车停车侵占。

The specified bikeway is occupied by illegal parking of motors in Wuhan.

广州某绿道，尽管绿道上人行道和自行车道在空间上完全分离，由于人行道被机动车停车占用，行人选择在自行车道上行走，造成事实上的未分离。这在绿道设计时应该予以充分的考虑。

Although the sidewalk and bikeway on the greenway in Guangzhou are completely separated in space, actually they are not separated as the sidewalk is occupied by illegal parking. This situation needs to be taken into fully consideration when designing greenways.

广州二沙岛上绿道自行车道，为规避前方占用自行车道停车的车辆，自行车骑出专用自行车道外，这无疑降低了绿道骑行的安全性。

The bikeway at Ersha Island in Guangzhou. Cyclists have to ride outside the special bikeway to evade illegal parking which undoubtedly lower the riding safety.

Chapter Two

东莞某绿道空间的划分未起到实际的意义

The spatial division of greenway in Dongguan doesn't make any actual sense.

宜昌某绿道与机动车道隔离

The greenway is separated with motor way in Yichang.

惠州某绿道在划分绿道空间时，应注重行人和自行车特有的出行属性，行人倾向于在道路最内侧行走，而此处的空间划分却未注意到这一点，实际的空间使用情况与设计相反。

Huizhou greenway. Travel features are crucial when designing the section of greenway. Pedestrians prefer to walk at the fringe of roads. This example shows that people are using the greenway in an opposite way of the design.

东莞某绿道在缺乏物理隔离设施的情况下，自行车道经常被机动车停车所占用。

The bikeway in Dongguan is often occupied by motor parking under the circumstance of lacking of physical separation facilities.

现象与设计方法
Phenomenon and Design Method

具体的分隔形式可依据绿道的位置、周边道路的等级、空间条件等选用不同的设计方法，如在道路较宽、车速较快的路段，应设置物理隔离；而在车速有限制的支路，自行车可采用颜色标线划定；具备公交站点的路段，要处理好自行车道与公交站的位置关系，减少进出站的公交车对自行车的干扰。

Specific separation form bases on greenways location, roads level and spatial conditions choosing different design approaches that setting up physical separation on wide roads with faster motor speed, using colored lines on branches with limited motor speed, properly dealing with the positional relation between bikeway and bus stops to reduce obstruction toward cyclists.

东莞某绿道隔离有效防止了机动车道的占道。

Greenway segregation effectively prevents the occupation of motor way in Dongguan.

阿姆斯特丹某绿道空间的清晰隔离保障了机动车、自行车、行人之间的相互独立

Clear separation of space ensures the mutual independence among motor vehicles, bicycles and pedestrians in Amsterdam.

广州中山大道绿道，通过不同高度的设置，分隔自行车道和人行道。

Zhongshan Avenue greenway has bikeway and sidewalk separated by setting of different altitudes.

Chapter Two

阿姆斯特丹某绿道空间的清晰隔离保障了机动车、自行车、行人之间的相互独立。

Clear separation of space ensures the mutual independence among motor vehicles, bicycles and pedestrians in Amsterdam.

乌得勒支自行车道设置在公交站台后,保障了机动车、自行车、行人之间的不干扰。

In Utrecht, ensure the non-interference among motor vehicles, bicycles and pedestrians by setting up bikeway behind bus stops.

巴塞罗那利用路边停车充当隔离,保障自行车道不受机动车干扰。

In Barcelona, use curb parking as separator to ensure that bikeway is not obstructed by vehicles.

现象与设计方法
Phenomenon and Design Method

阿姆斯特丹利用路边停车充当隔离，保障自行车道不受机动车干扰。

In Amsterdam, on-street parking is used to separate bikelane and vehicle lane, which ensures bikelane is not occupied by vehicles.

纽约利用公共自行车充当分隔带，保障自行车道不受机动车干扰。

In New York, use public bikes as separator to ensure that bikeway is not obstructed by vehicles.

哥伦比亚波哥大将某绿道设置在路中，保证绿道的独立性。

The greenway is set in the middle of the road to ensure its independence in Bogota, Columbia.

037

Chapter Two

1.4 绿道的可达
The Accessibility of Greenway

绿道不仅需要高质量的绿道空间，同样需要良好的可达性，确保使用人群可以方便、安全的到达和离开绿道。但是在国内目前的绿道规划与设计中，并未得到应有的重视。

Greenways require not only high quality greenway space, but also good accessibility to make sure users reach and leave greenways conveniently and safely. But during the existing plan and design of greenways in China, it is not gotten as much attention as it should.

宜昌某绿道建设了良好的绿道空间，但使用者到达绿道的通道处建设却被忽略。

In Yichang, good greenway space has been planned, but the construction of access road is ignored.

在广州对面即为临江绿道，却无法到达。

In Guangzhou, users cannot reach even the riverside greenway is in the opposite side.

广州萝岗某绿道，交叉口另一侧即为绿道，却无法直接连接。

Luogang greenway is situated at the other side of this intersection, but there is no direct connection for users walking towards there.

现象与设计方法
Phenomenon and Design Method

保障绿道的可达性，需要增设连接通道，同时，也需要确保使用者使用通道的安全性，路中安全岛能保证绿道中行人和非机动车的过街安全，为绿道路中等待的行人和非机动车提供保护，极大地提高行人和非机动车过街的安全性。抬高过街通道可有效降低机动车在通过过街通道时的车速，保障行人和自行车安全。

Connecting passageway is added to ensure greenway accessibility and make sure that users are safe when using it. Refuge island can ensure the safety of pedestrians and non-motorized transport in the greenway when they cross the street to provide them protection at the same time greatly enhance their crossing safety. The speed of motor vehicles when passing passageway can be lowered effectively through lifting intersections to ensure the safety of pedestrians and cyclists.

铜锣湾崇光百货外人流巨大，斑马线宽度也相应增加。（香港）

In Hong Kong, there are huge numbers of people outside the Tongluowan SOGO Department Store where zebra crossings are widened correspondingly.

在学校门外设置安全的过街设施，确保学生的安全。（香港）

In Hong Kong, safe crossing facilities are set out side the school entrance to ensure students' safety.

在斑马线前数米设置减速带，路中设置安全岛和灯光指示。（香港）

In Hong Kong, deceleration strips are set meters before zebra crossings and the safety island and indicator lights are set in the middle of the road.

原来的机动车道较宽，所以加设安全岛不会妨碍机动车的通行。安全岛两端抬升，并装有照明护柱，有效预防机动车的冲撞，斑马线在同一个平面上，与人行道相接处有无障碍坡道，通行方便安全。（香港）

The original motor way is wide enough for an additional safety island without obstructing the passage of motor vehicles. Both sides of the safety island are lifted with luminous bollard to prevent vehicles crash effectively, with zebra crossings in the same plane connecting sidewalks with barrier free ramps to make crossing convenient and safe.

Chapter Two

即使是较窄的路面,也需要安全岛给行人安全等候,尤其是不设信号灯的过街。(香港)

Safety island is also required at a relatively narrow street for people when they are waiting to cross, especially when the signal lamp is not planned. (in Hong Kong)

无论有无信号灯,都应在较宽的马路上设置行人安全岛。(荷兰)

Intersections reaching a certain width should be set with safety islands which should be installed at relatively wide streets no matter there is or isn't signal lamp. (in the Netherlands)

行人过街处地面抬升,上下坡处划有长短线提醒司机,在与人行道接驳处细心考虑无障碍通行,并配有限速标识。(阿姆斯特丹)

Lifted pedestrian crossing warns the drivers with lines at the edge and considers carefully the connection with sidewalks using barrier-free accessibility equipped with speed limit sign. (in Amsterdam)

行人过街处地面抬升,并涂上红色;把人行道往外扩,缩短过街距离;在人行道边缘装上护柱,防止机动车驶进。(巴黎)

The lifted pedestrian crossing is painted in red with extended sidewalk to reduce crossing distance and the edges of sidewalk are equipped with bollard to prevent motor vehicles from entering. (in Paris)

现象与设计方法
Phenomenon and Design Method

绿道不仅应串联起重要的城市景点，良好的公交服务也是提升绿道吸引力和扩大绿道辐射范围的重要因素，绿道应与公交系统紧密结合，方便人们轻松、愉快地到达公共交通站点，为市民使用绿道提供条件；同时，绿道提供的新的自行车道和人行道的高品质联接将会对现有城市交通网络起到补充与扩展的作用。

Greenways should not only connect important attraction points, good bus service is also an important factor to improve the attraction and radiation range of greenways which integrate closely with public traffic system to lead people to the public transportation junction easily and enjoyably; meanwhile the high quality connection of bikeway and sidewalk provided by greenways can supplement and extend the existing urban trip network.

地铁出入口直接连接绿道。

The subway entrance is connected directly with the greenway.

哥伦比亚波哥大某绿道和公交系统几乎无缝衔接。

The greenway and public transport system is connected seamlessly in Bogota, Columbia.

广州中山大道绿道与BRT、公共自行车整合到一起，为绿道提供了极为便捷的公交服务。

Zhongshan Avenue greenway is integrated with BRT and public bikes to provide extremely convenient bus service.

布里斯班某绿道、有轨电车站点、常规公交、自行车停车位都有机地整合到了一起，为使用绿道提供了方便。

In Brisbane, greenways, tram stations and bike parking are integrated together to provide convenient use to greenways.

1.5 绿道的服务
The Service of Greenway

步行和骑行对环境的舒适度较为敏感。能够吸引人们去享用绿道，必须是有一个宜人的小气候。平顺的路面、避免日晒雨淋，以及考虑周全的配套设施与服务都能大大提高绿道的舒适性及趣味性。在绿道服务上，国内目前通常是在绿道上设置驿站，提供必要的服务。但是，有限的驿站却很难为使用者提供多层次的服务。

Walking and cycling are relatively sensitive to environmental comfort. The condition that attracts people to enjoy the greenway must be to create a pleasant microclimate. All the factors including smooth road surface, shelter avoiding sun and rain, as well as thoughtful supporting facilities and services, can greatly improve the comfort and enjoyment of greenways. To the greenway service, normally courier station can provide necessary service, but limited courier stations are hard to offer multi-level services.

越秀区发展公园绿道驿站，并非所有时段都有服务人员值守。

Greenway courier station of development park in Yuexiu District, service staff is not on duty all the period.

宜昌某绿道驿站，驿站可提供的服务有限。

The greenway courier station in Yichang can only provide limited service.

现象与设计方法
Phenomenon and Design Method

事实上，绿道可为社会互动提供良好的场所。它的对象应包含所有的使用人群——当地居民和游客。绿道为周边社区带来巨大的客流，促进社区的经济发展，同时，社区的繁荣也可给绿道带来了更多的使用者。因此，绿道与周边社区的地面层应保持通行，为绿道与周边建筑的互动创造条件。

In fact, greenways provide ideal places for social interaction whose objects include all the users——local residents and tourists. Greenways attract huge number of people for surrounding communities to promote their development; in return prosperity of communities brings more users to greenways. Therefore, the ground floor of surrounding communities is advised to keep clear to create conditions for interaction between greenways and surrounding buildings.

纽约，某绿道沿线设置有休闲座椅和商店，供人们在游玩绿道时休憩和提供人与人之间交流和沟通的平台。

In New York, leisure seats and shops are set along the greenway making it as a platform for people to communicate with each other during their play and rest.

乌得勒支某绿道沿线设置有自行车停车位、休闲座椅和商店，供人们在游玩绿道时休憩和提供人与人互之间的交流和沟通的平台。

In Utrecht, bicycle parking, leisure seats and shops are set along the greenway making it as a platform for people to communicate with each other during their play and rest.

广州某绿道沿线的体育设施吸引更多市民使用绿道。

In Guangzhou, sports facilities along the greenway attract citizens to use the greenway.

Chapter Two

其二，中国南方许多城市高温多雨，但绿道上往往缺乏乔木或风雨廊遮阳挡雨，使行人常常暴露在不舒适的天气中。同时，绿道充足的照明也是夜间正常使用绿道的最基本保障。

Secondly, many cities in south China suffer high temperature and rainy days, but there are often short of plants or shelters in greenways forcing pedestrians to expose themselves to the uncomfortable weather. At the same time, sufficient lighting is also the most basic security for using greenways at night.

在布里斯班，专用绿道桥梁，不仅提供了遮盖，也设置了驻足休憩的区域，而且也提供了充足的照明设施，绿道系统在夜间也能方便地使用。

In Brisbane, the special greenway bridge not only provides a shelter, but also sets up a resting area with sufficient lighting for night using.

广州某绿道提供良好的照明，旁边设置休闲座椅，吸引市民在绿道上停留。

Equipped with leisure seats, the greenway with good lighting attracts people to stay on it. (in Guangzhou)

马尼拉某绿道提供遮阳设施，确保使用者能够舒适地使用绿道。

In Manila, the greenway is equipped with shading facilities to make sure that users enjoy the greenway comfortably.

广州某绿道良好的照明是夜间能够得到正常使用的基本保证。

Sufficient lighting is the most basic security for the greenway to be used at night. (in Guangzhou)

现象与设计方法
Phenomenon and Design Method

其三，国内城市普遍缺乏高质量的自行车停放设施，自行车停放的安全性成为骑行者的负担，降低了骑行的积极性。

Thirdly, cities in China are usually lack of high quality bicycle docking facilities making bicycle docking safety the misgivings to the cyclists thus reduce their riding positivity.

其四，绿道指引系统并不完善，未能提供使用者足够的指引信息服务。

Fourthly, the indicating systems are far from perfect without providing users enough indicating information service.

布里斯班某绿道自行车停车点，不但设置有U型停车架，自行车停车点也有遮盖和照明设施。

In Brisbane, the greenway bicycle parking is equipped with not only U-shaped racks but also shelters and lightings.

华盛顿中央车站的自行车停车点和租赁点。

The bicycle parking and rental station in the central station. (in Washington)

广州部分区段绿道不连续，在绿道的尽端，也缺少必要的指引信息或地图指引使用者。

The greenway in Guangzhou is discontinuous, and there is seldom necessary information or map at the ending point to guide the users.

阿姆斯特丹，专用的自行车信号灯。

The special bikeway signal lamp in Amsterdam

045

Chapter Two

布鲁塞尔某绿道周边设置有公共自行车，公共自行车站点还设置了详细的地图。

In Brussels, public bicycles are set around the greenway with detailed maps in the public bike stations.

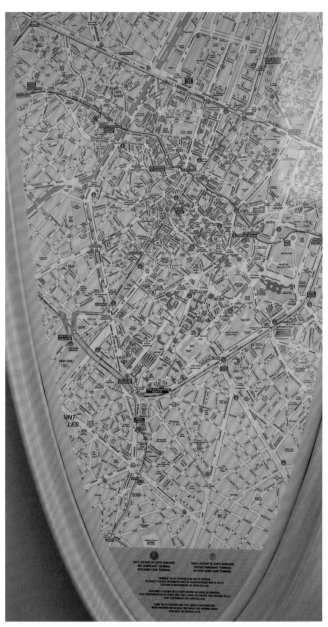

鹿特丹每条绿道都有自己独有的编号，更好的指引使用者。

In Rotterdam, each greenway has its unique number providing users a better guidance.

布鲁塞尔某信息指引地图相当的而详细。

The indicating map is quite detailed. (in Brussels)

2 绿道设计方法
The Design Method of Greenway

中国绿道在使用中存在的"连通、连续、隔离、可达、服务"等五个问题，是完全可以通过技术手段进行解决的。国外城市在绿道建设时，也曾经有着同样的情况，目前，已经有了较为成熟的解决办法。这些设计手法实施所需花费相对便宜，可以方便地实施，在中国绿道的规划、建设中起到重大作用。

In China, the five issues, "connectivity, continuity, separation, accessibility and service", existing in the use of greenways can be solved sufficiently through technological methods. When greenways were constructed in foreign cities, there were some similar situations which can be solved by relatively mature solutions now. These design methods can play a major role in greenway planning and construction in China with relatively low cost and convenient implementation.

2.1 网络规划
Network Planning

2.1.1 密集而完整的网络
Dense and Complete Network

现在中国典型的城市新区开发以 400~500 米的大街区为主，街区以宽阔的主干道为边界，各个街区相互孤立，难以形成密集的非机动车交通网络，大大增加了步行与骑行的绕行距离，这也给绿道的网络设计带来不便，中国绿道之间常常会出现因大型交通设施、地理因素而无法互通，所以无法形成真正的绿道网络，降低了绿道的吸引力。规划密集的绿道网，可以减少绕行距离，增加路径的多样性，使步行与骑行更便捷、舒适、有趣。小街区的划分有助于形成密集的绿道网络。开放部分社区慢行网络，作为城市绿道的重要补充。

At present, Chinese typical urban new area development is mainly large block of 400–500 meters with the edges of which are wide main streets making each block isolated with each other. It is difficult to form a dense non-motor traffic network, increasing greatly the detour distance of walking and riding, causing inconvenience to the design of greenway network. Because of large-scale traffic facilities or geographic factors, each greenway cannot be interlaced together to form a real network thus reduce greenway attraction. Planning dense greenway network is able to reduce detour distance at the same time increase the route diversity to make walking and riding more convenient, comfortable and interesting. The division of small blocks contributes to form a dense greenway network. Open a part of community non-motorized transport network as an important supplementation of urban greenway.

Chapter Two

为提高绿道可达性，我们可以引入绕行系数这一评价指标，评价绿道的可达性。绕行系数是指起始点与目的地之间的实际出行距离与直线距离的比，绕行系数越低，则服务水平越高。在绕行系数较高的区域（绕行系数在 D 以上），需要通过增加行人和自行车通行的通道，来提高绿道的可达性。

In order to improve the accessibility, we can introduce the evaluation index of detour factor to evaluate the accessibility of greenway. Detour factor refers to the ratio between the actual trip distance and the measured distance between the start point and the destination with the lower detour factor corresponding to the higher level of service. At areas with relatively high detour factor (higher than D), more sidewalk and bikeway are required to improve the accessibility of the greenway.

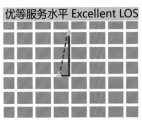

优等服务水平 Excellent LOS　　中等服务水平 Medium LOS　　较低服务水平 Low LOS

- - - - 测量距离 Measured Distance
—— 实际距离 Actual Distance
▲ 起讫点 Origin and Destination

· 服务水平（LOS）：通过等级评分（A 到 F）评价该地区居民到目的地的方便性及直达性。

· Level of Service (LOS): The Rates (A to F) to evaluate the convenience and directness of walking to destinations.

服务水平 Level of Service	绕行系数 Detour Factor
A	<1.2
B	1.2-1.4
C	1.4-1.6
D	1.6-1.8
E	1.8-2.0
F	>2.0

· 绕行指数：实际行走距离与直接测量距离之比（实际行走距离 / 直接测量距离）。

· Detour Factor: The ratio between the actual pedestrian distance and the measured distance between the BRT station and a selected site.(Actual distance / Measured distance)

绕行系数评价指标图
Detour factor evaluation indicatrix

广州六运小区

Guangzhou Liuyun Community

广州六运小区北临中山大道绿道，南接花城广场绿道。原为封闭居民小区，2009 年，小区进行大规模改造，整个小区被强化为行人主导的空间，禁止机动车通行的道路把机动车在社区环境的影响降到最低，同时为绿化和步行区域留出了更多的空间。

Guangzhou Liuyun Community was originally a closed residential area with Zhongshan Avenue greenway to its north and Huacheng Square greenway to its south. In 2009 after a large-scale redevelopment, the community had been improved into pedestrian dominated space with the non-motor roads that minimized the impact of vehicles towards the community environment at the same time leaved more space for greening and walking.

现象与设计方法
Phenomenon and Design Method

049

2.1.2 便利的公共交通服务
Convenient Public Transport Service

在规划绿道网络时，应考虑到公交因素，尽量保证绿道网络有良好的公交可达性。

Public transport factor is required to be taken into consideration trying to ensure the greenway network that provide good accessibility for public transportation..

广州中山大道 BRT 走廊沿线的绿道。
The greenway along the line of the Zhongshan Avenue BRT (in Guangzhou).

现象与设计方法
Phenomenon and Design Method

波哥大旧城中心的公交步行街。
The Transit Mall at the old city centre of Bogota.

中山大道绿道旁的地铁石牌桥出入口
Shipaiqiao Metro Station Entrance beside Zhongshan Avenue Greenway

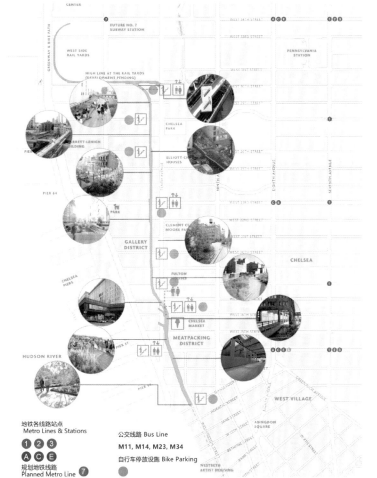

纽约高线公园周边有多条地铁线路到达，并且有多个公交站点。
There are many metro lines and bus stations around the New York High Line Park.

2.2 基础设施设计
Infrastructure Design

要吸引更多的人使用绿道，首先需要建设良好的绿道。对于使用者来说，高质量的绿道，首先要拥有高质量的路面，充足的使用空间，无障碍物阻隔，以行人、自行车优先的设计并且便利的到达性和使用者的安全性。

To attract more people to use greenways, the first step is to construct a good greenway. For users, a high quality greenway should possess high quality road pavement, sufficient space, barrier-free design, design prioritize pedestrians and cyclists, and provide convenient accessibility as well as users' safety.

2.2.1 交叉口与路中过街
Intersections and Mid-Block Crossing

为了确保绿道在与机动车道路交汇时的连续性，又保障行人和自行车的安全，交叉口与路中过街通道的设计应达到如下效果：
降低机动车穿过绿道的车速；
规范机动车行车路径；
保障自行车与行人的专用路权和路权的优先级别；
为行人与自行车提供安全庇护；
缩短行人过街距离；
强化对机动车的提示作用。
这些设计目标，可通过以下设计方法实现。

In order to insure the continuity when the greenway intersects with the motorway, as well as the safety of pedestrians and cyclists, the design of intersections and crossings is suggested to achieve the following effects:
Slow down the speed of motor vehicles when passing through the greenway;
Regularize the route of motor vehicles;
Ensure the special road right of cyclists and pedestrians, as well as the priority level of the road right;
Provide protection for pedestrians and cyclists;
Shorten the crossing distance to pedestrians;
Strengthen the effect of hints to motor vehicles.
These design objectives can be achieved through the following design methods.

1) 设置过街安全岛

Refuge Island

安全岛两端应突出作为防护，两侧人行道应设置无障碍坡道，为防止机动车驶入人行道，坡道需加装护柱。

Both ends of the safety island extrude as protection and both barrier-free ramps with bollards should be set at both sides of the sidewalk to prevent motor vehicles driving into.

并非较宽的道路才需要设置安全岛，在双向道路中，如无信号灯控制，宜尽量压缩两侧车道，设置安全岛。

Not just being installed at relatively wide roads, the safety island is advised to be set at two-way road that has both sides of the lanes compressed if there is no signal lamp.

Chapter Two

BRT车站为过街通道提供安全岛。（广州中山大道）

The BRT station acts as a safety island to the crossing passage. (at Zhongshan Avenue in Guangzhou)

便利行人往来的街道两侧可以进行商业活动，因此，在商业发达的非交通主干道中，可设置连续的路中安全岛，供行人在任意位置过街驻足。（伦敦牛津街，伊斯坦布尔）

In order to facilitate pedestrians, both sides of the street are planned with commercial activities, therefore at the non-primary route with developed commerce, continuous central refuges can be set for pedestrians to cross or stop at any place. (London Oxford Street & Istanbul)

2) 自行车专用过街通道

Dedicated Bike Crossing

在交叉口和路中过街应提供自行车的专用过街通道，既能确保自行车通行的连续性，又能保证自行车与行人各行其道，不互相干扰。

Special passage for bicycles is suggested to be planned at the intersection or in the middle of the road to ensure not only the continuity of riding but also cyclists and pedestrians follow their own lanes without mutual interference.

自行车过街也需要安全岛的保护，安全岛应有足够的宽度容纳自行车。

Bicycles also need to be protected when crossing the street, so the safety island should wide enough for a bicycle.

Chapter Two

可在机动车停车线前设置自行车等候区,在信号灯转换时,自行车可以提早起步避免混入机动车车流。

Bicycle waiting area can be set in front of the motor stop lines where bicycles can start earlier to avoid the mixing with vehicle stream when the signal light switching.

自行车过街廊道应有鲜艳的颜色或清晰的标线,以增强其对机动车的提示作用。

The street crossing for bicycles is required to be painted with bright colors or clear lines to improve its warning effect for motor vehicles.

3) 抬升的过街通道

Raised Crossing

抬升的行人过街通道可以降低机动车车速，并使路面保持平顺，而且与机动车道不一致的铺装（人行道铺装或鲜艳的涂料），不但可以提高其对机动车的提示作用，也可以降低机动车的车速。

The lifted street crossing is able to lower the motor speed and keep the road surface smooth. With its different pavement (painted with bright-colored coating), the lifted street crossing can not only improve the warning effect to the vehicles, but also lower the vehicle speed.

交叉口的整体抬升适用于优先非机动车交通的街道，如历史城区、居住社区、商业区或校园。

The entire elevation of the intersection is appropriate for streets of non-motorized transport prioritization, such as historic urban area, residential community, commercial area or campus.

由于路面抬升后，机动车容易驶入人行道，所以在人行道边缘需设置护柱。

Bollards need to be installed at the edge of the sidewalk as motor vehicles can enter easily after the road surface being lifted.

Chapter Two

4) 外拓人行道

Extend the sidewalk

在有路边停车的路段的行人过街通道处，人行道边缘可往外扩展至车行道边缘，从而缩短行人过街的距离，提高安全性，又能提供更多的步行驻足空间。

At the pedestrian crossing of section with curb parking, the edge of the sidewalk is suggested to be extented outside to the edge of the road to reduce the distance of street crossing for safety at the same time to provide more stopping spaces for pedestrians.

在道路改造中，如需降低工程量，使道路空间更为灵活，可用护柱界定车行道边缘。

Use bollard to define the edge of the road to lower the quantities and make the road space more flexible during the road reconstruction.

现象与设计方法
Phenomenon and Design Method

5) 信号灯

Signal lamp

在绿道过街时，可设置专用的自行车和行人过街信号灯，并通过信号灯相位的设置，保障绿道自行车与行人的优先等级。

Special signal lamp for bicycles and pedestrians can be installed at the crossing street to ensure their priority through the setting of signal lamp phase.

2.2.2 提升机动车出入口
Raised Vehicle Access

因为机动车出入口会频繁穿越绿道,所以为保障绿道的连续性,对绿道上机动车出入口,建议采用抬升地面的设计方式,保障绿道的平顺性。抬升后的出入口地面因为坡度提高,可起到降低车速的作用,保障行人及自行车出行的便利和安全。由于路面抬升,机动车容易驶入人行道,所以在人行道边缘需设置护柱。

The vehicle access on the greenway is suggested to use lifted design to ensure the continuity and smoothness as the greenway will be cut off by the access frequently. Because of the gradient, the lifted access lower the vehicle speed to ensure the convenience and safety for pedestrians and cyclists. Bollards are required to be set at the edge of the sidewalk as vehicles are easy to enter after the road surface being lifted.

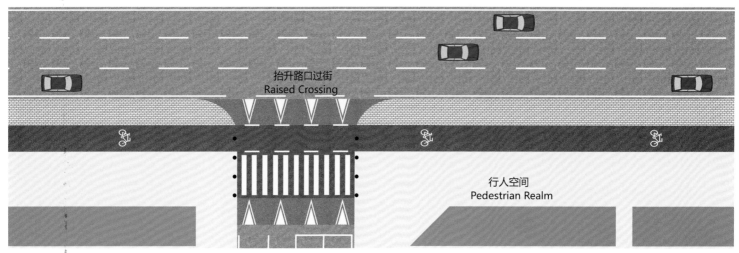

建议机动车出入口的设计
The suggested design of vehicle access

2.2.3 绿桥与隧道
Greenway Bridge and Tunnel

在有些情况下，绿道可能无法在地面保持连通，这就需要设置行人和自行车专用的桥梁或隧道来解决连通的问题。绿道桥梁可以连接河流或高等级道路路两侧的绿道，绿道桥梁也能很精美，很多城市的绿道桥梁已成为城市的地标。如伦敦泰晤士河上的千禧桥，已成为该城市靓丽的风景线。

Under some circumstances, the greenway may not solve the problem through ground connection so this is the reason why special bridges or tunnels for pedestrians and bicycles are required to be set. Rivers or greenways on both sides of the high-grade road can be connected by a greenway bridge which has become an urban landmark in many cities with its exquisite appearance. For example, the Millennium Bridge on the Thames has become a beautiful scenery of the city.

设计目标：缩短自行车与行人的出行距离；保障自行车与行人的专用路权。

Design objective: to shorten the trip distance of cyclists and pedestrians; to ensure the exclusive right-of-way of cyclists and pedestrians.

绿桥
Greenway bridge

Chapter Two

英国纽卡斯尔的某绿桥采用旋转的形式,在船只通过时,绿道禁止通行,并旋转至足够的通航高度。

The greenway bridge is adopted with revolving form. The bridge revolves to an enough height for navigation with pedestrians and cyclists banned from passing when ships get through. (in Newcastle, the UK)

桥上应清晰划定出自行车与行人的通行空间。

The passing space for cyclists and pedestrians should be clearly delimited on the bridge.

在引桥出入口需要有清晰的标识及护柱等防止机动车进入。

Clear signs and bollards are required at the approach bridge access to prevent vehicles from entering.

现象与设计方法
Phenomenon and Design Method

隧道也应设置人行道与自行车道,以及提供足够的照明。

The tunnel is also advised to provide sidewalk, bikeway and sufficient lighting.

绿桥需提供遮阳棚和足够的照明。

The greenway bridge is suggested to provide sunshade and sufficient lighting.

2.2.4 功能区分隔
Division of Functional Zone

为了减少机动车对绿道的干扰，从而提高绿道的安全性及舒适性，需要对机动车道和绿道进行清晰的区域界定，而自行车和行人在使用绿道的时候，行为习惯和体验也有所不同。因而，确保绿道安全、有序，关键在于处理机动车、自行车与行人之间的冲突，其最重要的措施是实施明确的功能分区，保证三者各行其道，从而减少冲突。

In order to reduce the interference of vehicles and improve the safety as well as convenience, the areas of motorway and greenway are required to be defined clearly. The behavior habits and experiences are different between cyclists and pedestrians when using the greenway. Therefore, the key that ensures a safe and ordered greenway is to deal with the conflict among motor vehicles, cyclists and pedestrians, especially implementing a specific function division to ensure they go their own ways thus reducing conflict.

设计目标：保障自行车与行人的专用路权；为行人与自行车提供庇护；加强绿道对机动车的提示作用。

Design objective: to ensure the exclusive right-of-way of cyclists and pedestrians; to provide shelter for pedestrians and cyclists; to enhance warning effect that the greenway presses to motor vehicles.

功能区分隔有以下几种方式。

The several ways of division of functional zone are as follow.

1) 物理隔离

Physical Separation

(1) 护柱

Bollard

通过设置隔离护柱分隔绿道和机动车道，是目前发达国家运用最为普遍的方式。

The most common way that developed countries use at present is to separate the greenway and motorway by setting bollards.

绿道出入口位置也需设置护柱，防止机动车驶入。

Bollards are also installed at the greenway access to prevent vehicles from entering.

(2) 道路缘石
Curb

通过设置隔离缘石分隔绿道和机动车道，其作用与护柱类似，也是目前普遍运用的方式。

Setting up curbs to separate the greenway and motorway is also the common way used at present similar with bollards.

通过不同的高差处理分隔绿道中的人行道与自行车道。

Separate the sidewalk and bikeway in the greenway by different altitude.

(3) 公共自行车与路边停车等市政设施隔离
Separate by using public facilities like public bikes and curb parking

路边停车作为自行车道与机动车道的分隔时，需要预留一定的空间，防止机动车开门时影响自行车通行。

In order not to impact the bicycle transit, reserve certain spaces for vehicles opening doors when the curb parking acts as the separation between the bikeway and the motorway.

通过使用公共自行车和路边停车设施等市政设施隔离，主要运用在城市建成区的，通过对旧有的城市空间进行重新安排，在赋予城市空间新的功能、调整城市空间用途时，实现了对城市空间的充分利用。

The way of separation by public bikes or curb parking is mainly used in the urban built-up area that achieves a sufficient use of the urban space at the same time endows it new function and adjusts its purpose through the rearrangement to the old urban space.

2) 颜色标线
Colored lines

在宽度较窄、车速较低的单行路段，可设置反向的自行车道，同向的自行车可与机动车混行。

At the relatively narrow one-way road segment with low speed, set a reverse bikeway while bicycles of same direction mix with motor vehicles.

3) 公交站点的路段
Road segment with bus stations

自行车道应设置在公交站与人行道之间，避免公交车进出站对绿道的干扰。

The bikeway is suggested to be planned between the bus station and the sidewalk to avoid the interference to the greenway when buses drive in and out to the station.

现象与设计方法
Phenomenon and Design Method

4) 非机动车优先的共享街道
Non-motorized traffic prioritized shared street

在部分城市支路、社区道路中，如非机动车交通为主导，又需要保留少量机动车通行的需求，可以考虑设计成行人和自行车优先的共享街道。共享街道的设计是上述设计手法的综合运用，并利用蜿蜒的线型降低车速，结合丰富的景观设计和街道家具的设置，提升步行和骑行的体验。

A part of urban branches and community roads are suggested to be designed as shared streets of pedestrians and bicycles priority if non-motorized traffic is dominant at the same time need to keep the access of . The design of shared street is the comprehensive application of the design method mentioned above. Use winding line to slow down the speed, combining more plentiful landscape design and the setting of street furniture to improve the experiences of walking and riding.

较窄的车道可以降低机动车车速。
The relatively narrow lane can slow down the speed.

蜿蜒的线型，同样也能起到降低机动车车速的功能。
The winding line can also slow down the speed of vehicles.

Chapter Two

虽然街道没有高低分隔，行人行走会更加平顺，但需要在机动车道边设置护柱加以限制。在道路空间内，机动车只能在限定的区域内行驶，而整条街道自行车和行人均可不受限制地使用，优先级别高于机动车。

Pedestrians walk more smoothly on the street without altitude difference, but bollards are required to be installed at the side of the motorway for restriction. Motor vehicles are only allowed to drive in the restrictive area in the road space while the priority level of pedestrians and bicycles is higher than vehicles in the whole street without any restriction.

只允许公交车、自行车与行人通行的公交步行街适合用于道路空间有限的历史城区和商业区。

The Transit Mall that allows buses, bicycles and pedestrians to enter only is suitable for historic urban area and commercial area with limited road space.

2.3 配套设施设置
Supporting Facility Setting

好的绿道系统，应更多地方便市民的参与和使用。这就需要满足市民在使用绿道时衍生出来的其他的需求，并为此在合适的位置设置配套设施。

A good greenway system should be more convenient for citizens to participate and use. This requires to satisfy the other needs which are derived from the moment when citizens use the greenway therefore to set supporting facilities in place.

2.3.1 风雨廊
Covered Walkway

风雨廊连接重要的公共建筑、住宅区与公交站点。可在较恶劣的外部环境下营造出怡人的小环境。适用于使用人群较多的重点区域。

The rain shelter that connects important public buildings, residential areas and bus stations can create a pleasant environment in a relatively severe external environment. It is suitable to be applied in key areas with more users.

风雨廊与商业设施结合。
The rain shelter integrates with commercial facilities.

风雨廊与座椅等街道家具结合。
The rain shelter integrates with street furniture like seats.

风雨廊的设计形式多样，可以简洁而低成本，也可以具有地方特色的同时提供足够的照明。

The shelter can be designed in various style. It can have simple structure with lower the cost. It can provide sufficient lighting. It can also exhibits the local characteristics.

2.3.2 自行车停放设施
Bicycle Parking Facility

能够让人们在绿道上骑自行车倒是一件令人愉快的事情。因而在绿道附近配套设置易于使用的自行车停放设施变得尤其重要。

It is delightful to let people ride their bicycles on the greenway. So it is particularly important for the convenient bicycle parking dock set around the greenway.

自行车停放设施的设置方式多种多样，可以在人行道装设简单的停车架，但尺寸应能满足锁上自行车前后轮或车架，存取车方便安全，停车架可设置在两侧供自行车停放，提高停车的效率。

Phenomenon and Design Method
现象与设计方法

There are various setting types for the bicycle parking facilities. Install a simple parking stand on the sidewalk with its size been able to lock with the wheels or the frame of the bike conveniently and safely. Bicycles can be parked at both sides of the stand to enhance the efficiency of the parking facility.

双层自行车架可以节省空间，停放更多的自行车。

The double-deck bicycle stand can save the space and park more bicycles.

可以利用行道树之间的空间设置自行车停车架。

Use spaces between border trees to set the bicycle stand.

在自行车停放点增加挡雨棚，避免自行车日晒雨淋。

Install the rain shelter for the bicycle parking dock to avoid the bikes from being exposed to the sun and rain.

空间较为充裕时，可设置自行车停车箱，增加安全性，同时可以避免自行车日晒雨淋。

When the space is ample, set up the bicycle parking box to increase safety at the same time avoid the bikes from being exposed to the sun and rain.

一个机动车车位能停放的自行车数量可达10辆以上，在空间有限的区域，用机动车车位作为自行车停放点是很好的解决方法。

In the area with limited space, using the vehicle parking spot as the bicycle parking dock is a good solution, as one vehicle parking spot is capable to park more than ten bicycles.

在停车需求大的地点，如公交站点、城际火车站等，可以设置自行车停车楼或地下自行车库，建筑可以成为城市的风景线。

At the place with large demand of parking such as bus station and intercity railway stations, set up an underground bicycle parking garage or a bicycle parking building which becomes an urban scenery with its appearance.

2.3.3 公共自行车
Public Bicycle

巴黎利用某路内停车位设置公共自行车

In Paris, the curb parking spot is transformed into public bicycle parking docks.

伦敦某公共自行车系统

The bicycle sharing system in London.

公共自行车系统在任意服务点"通租通还"的模式使人们不用担心自行车被偷盗的问题，也不必为寻找自行车停车点而苦恼，更为不具有自行车的个人及家庭提供了便利条件。同时，在公共交通站点、绿道、商务区及居住区设置公共自行车服务点，实现骑自行车从家到绿道、工作地点的连通，进一步延伸绿道的功能和内涵。

The mode that rents or returns at any point of the bicycle sharing system let people no need to worry about the bike being stolen or finding a parking dock, and provides convenience for those people or families without owning a bike. Moreover, achieve the connection to the greenway or to the workplace from home by bike through setting public bicycle service points at public transport stations, greenways, business districts and residential areas to extend the function and content of the greenway.

现象与设计方法
Phenomenon and Design Method

2.3.4 行人指引系统
Pedestrian Indicating System

行人指引系统为绿道使用者辨明方向、规划路径，因此需要准确、易读的信息。指引系统必须清楚解释周边环境，提供清晰和必需的信息，有可读性强的地图指引。而且区域内应有统一的系统，分级设计，并摆放在交叉口及显眼的地方。设计形象应有地方特色，还可结合二维码等新技术手段。

The pedestrian indicating system tells the direction and plans the route for the users with accurate and readable information. The system is required to explain the surrounding environment clearly at the same time provide distinct and necessary information with map guidance of strong readability. The area is required to install a unified system of hierarchic design which is placed at the intersection or prominent location. The design is suggested to possess local feature and combine new technologies like 2D barcode.

伦敦某行人信息指引系统

Pedestrian indicating system in London

Chapter Two

2.3.5 自行车计数器
Bicycle Counter

安装在自行车道的自行车计数器,在每一个骑行者经过的时候,所显示的数字就会加一,为经过的骑行者提供有趣的骑行体验,又能唤起社会大众及政府部门对自行车使用者的重视。

The displayed number of the bicycle counter installed at the bikeway will plus one when a cyclist passes to provide cyclists with interesting riding experience at the same time arouse the attention of the public and the government to bicycle users.

小结
Conclusion

本部分的关注点集中在绿道的技术细节之上,从绿道使用者的角度出发,以国内绿道目前普遍存在而又未引起充分重视的"连通、连续、隔离、可达、服务"五个问题入手,借鉴国际上绿道建设使用的实践经验,结合国内的实际情况,对中国绿道解决这些问题的具体技术细节进行探索,并使用国外成熟的设计手法,通过技术手段来解决中国绿道存在的问题。也期望这些可行的设计方法能在中国得到广泛的应用,并成为绿道设计的技术方法,在未来的绿道建设中有所帮助,提升中国绿道的吸引力和综合效益。同时,我们也认识到,高质量的绿道离不开考虑周全的设计,更离不开持久的精心维护与管养。绿道在城市中不仅发挥着生态功能,更承载着社会、交通、经济等复合功能,也希望高质量的绿道能在未来城市发展中起到越来越重要的作用,成为引导城市可持续发展的重要一环。

Focusing on the technical details of greenways, this part refers to the practical experiences of international greenway construction combining with the domestic practical situations to explore the specific technical details of these problems of greenways in China and uses foreign mature design methods to solve the existing problems through technical means from the perspective of greenway users, starting with the five problems of "connectivity, continuity, separation, accessibility and service" which are common but lack of enough attention. We also expect that these feasible design methods can be widely applied in China to become the technical methods of greenway design supporting the greenway construction in the future as well as improving the attraction and comprehensive benefits of greenways in China. Meanwhile, we realize that high quality greenways can not be achieved without thorough design or sustained elaborately maintenance and management. Greenways in cities not only exert the ecological function, but also carry the complex functions like society, transportation and economy. We hope that high quality greenways play a more and more important role in the city development in the future to become a significant step that leads the urban sustainable development.

第三章
Chapter Three

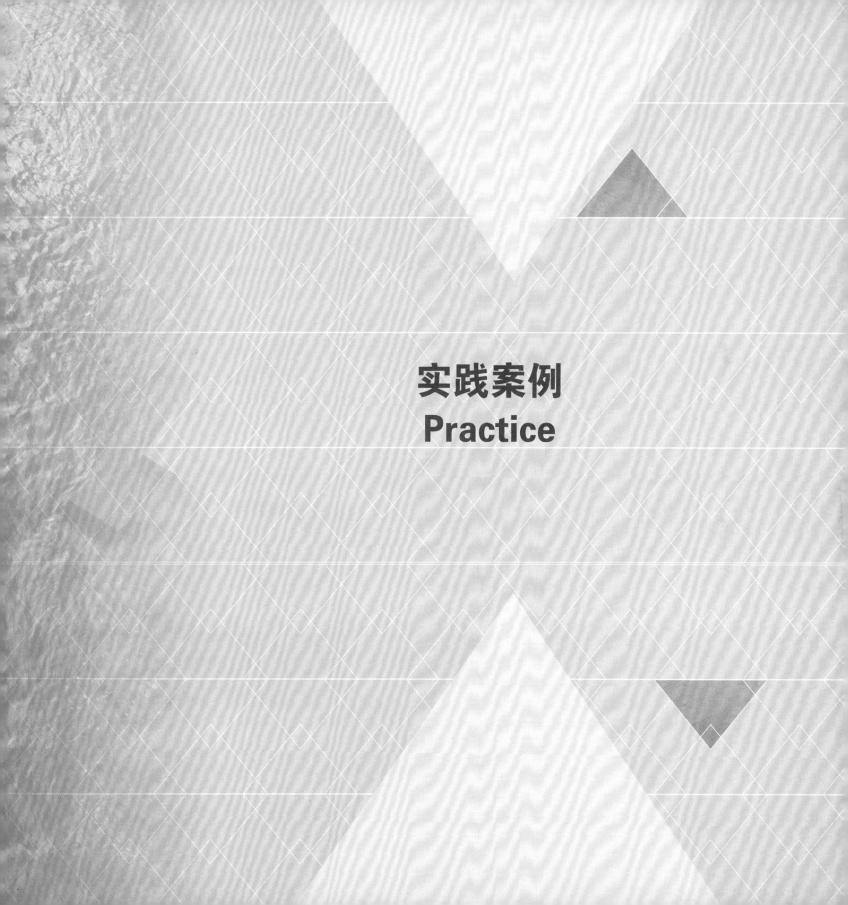

实践案例
Practice

Chapter Three

荔湾旧城慢行系统改善建议

Liwan Old Town NMT Improvement Suggestion

完善高质网络
To Optimize High Quality Network

1.1 人行道
Sidewalk

人行道综合评价是结合人行道实际宽度与道路障碍物作出判断，标准如下：

质量较好：宽度大于 3 米，无障碍物；

质量一般：宽度为 2~3 米，无障碍物；或者宽度大于 3 米，有少量障碍物；

质量较差：宽度小于 2 米；或者宽度为 2~3 米，有障碍物。

The sidewalk comprehensive evaluation focuses on the actual width and the obstacles. The standard is as follows:
Good quality: wider than 3 meters, no obstacle
Medium quality: 2~3 meters, no obstacle; or wider than 3 meters with a small amount of obstacles
Poor quality: less than 2 meters; or 1~2 meters with obstacles.

部分车辆违章停车，摊贩占道经营，使得人行道更狭窄。

Part of the vehicles' illegal parking and market stall's occupancy narrow the sidewalk.

荔湾旧城人行道的综合评价偏低，主要存在以下问题：

In the Liwan Old Town, the rating of comprehensive evaluation of sidewalk is relatively low with the following problems:

①人流量大的地方人行道宽度不足，行人被迫走到机动车道上。
②机动车违规停放，摊贩占道经营，以及多种市政线杆进一步缩小人行道的有效宽度。
③自行车通行空间得不到保障，常常出现自行车与行人抢用人行道的情况。
④部分人行道被小区和商场的停车场出入口车道打断，人行道不连续。
⑤无障碍通行设施不完善。
⑥缺乏清晰易懂的行人导向系统。

① The width of the sidewalks in the places with large pedestrian flow is not wide enough; pedestrians are forced to walk on the vehicle lanes.
② Illegal parking, market stalls and some infrastructure poles shorten the width of the sidewalks.
③ Cycling space is not ensured; bicycles often mix with pedestrians on the sidewalks of limited width.
④ The sidewalks are interrupted by the vehicle access of neighborhoods, shopping malls and parking lots.
⑤ There are many problems with the barrier free access facilities.
⑥ Lack of a clear and easily understandable pedestrian guidance system.

自行车道被装、卸货的车辆占用，自行车和行人抢用2米不到的人行道。

Bicycle lanes are occupied by unloading cargo, bicycles and pedestrians crowded in the sidewalk narrower than 2 meters.

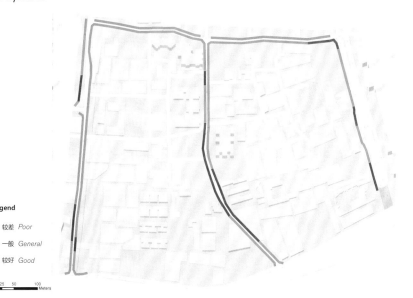

调研区域人行道综合评价

Comprehensive evaluation on the sidewalk in researched area

Chapter Three

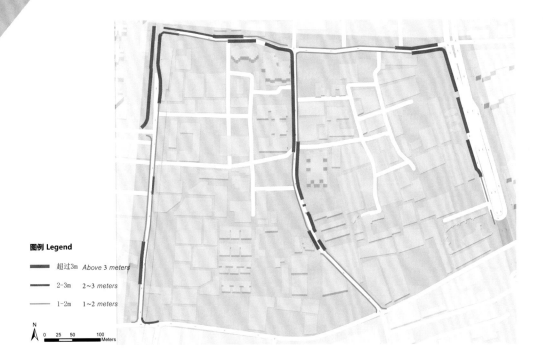

调研区域人行道宽度

The width of the sidewalk in researched area

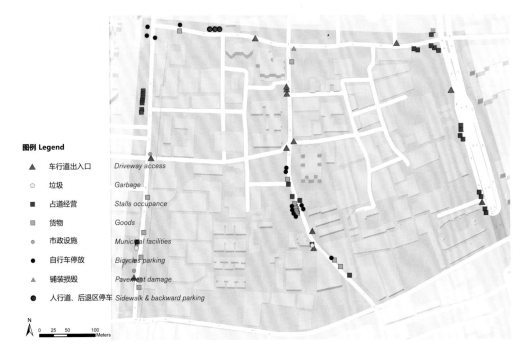

调研区域人行道障碍物

The obstacles in the sidewalk of researched area

实践案例
Practice

人行道被垃圾桶及三轮车占用。
Sidewalk is occupied by tricycle.

多辆自行车沿街停放。
Many bicycles park along the street.

车行道出入口频繁地打断人行道。
The entrance and exit interrupt sidewalk frequently

商铺占道经营,将货物堆放到人行道上。
Market stalls occupancy on the street, and put goods on sidewalk ..

081

Chapter Three

专题：创建连续的人行道

Topic: Build Continuous Sidewalk

抬升机动车出入口地面，使人行道路面连续，在路面用颜色及标志提示司机，在人行道两侧加护柱，防止机动车驶入人行道。充分体现行人优先的理念，使机动车转入时减速，提高行人的安全性。

Raise the vehicle access to ensure the continuity of the sidewalk; distinctive pavement colour reminds drivers to slow down. Bollards on both sides the sidewalk prevent vehicles from occupying the pedestrian areas. This fully represents the concept of pedestrian priority, by forcing vehicles to slow down and improving the safety of pedestrians.

快速穿过人行道的机动车威胁了众多行人的安全。
Vehicles' rapid turning threatens the safety of pedestrians.

实践案例
Practice

文昌路骏业阁外
Junyege, Wenchang Road

长寿西路文昌阁外
Wenchangge, Changshou Road. (w)

Chapter Three

方案：康王路华林国际东北角

Plan: Northeast Corner of Hualin International in Kangwang Road

康王路的行人数量与自行车流量非常大。行人和自行车应该具有通行的优先权，当机动车转入支路时应减速慢行。

The pedestrians and bicycle flow in Kangwang Road are very large. Pedestrians should have the right of first priority, while the vehicles on the road should slow down their speed.

建议抬升车行道出入口地面至与人行道同高，该段行人与自行车过街地面涂装红色以警示机动车，并增加慢行标识。

Entrance and exit of motor vehicle should be uplifted to the same height with sidewalk. The pedestrians crossing and bicycle lane should be marked in red to warn the motor vehicles and increase slow down signal.

1.2 自行车通行环境
Cycling

专题 1：自行车道

Topic 1: Bicycle Lane

现状问题 Current Situation

① 自行车流量大，但自行车道狭窄，某些路段只有0.5米宽；主干道康王路上的自行车道也偏窄，细节设计不合理，骑行者多选择使用机动车道，对他们自身及其他道路使用者造成安全隐患。
② 部分自行车流量大且机动车车速较高的干道没有设置自行车道。
③ 大部分主干道上的自行车道没有设置隔离设施。
④ 机动车违章停放，占用自行车道。

① Large bicycle flow but narrow bicycle lane. Some road section has only 0.5 meters width. The bicycle lane in Kangwang Road is also narrow. Some detail designs are not rational. Most of the bicycle rides on the motor vehicle lane. It will cause safety hazard for both bicycle riders and other users.
② Part of the main roads with large bicycle flow and high motor speed are not set with bicycle lane.
③ Most of the bicycle lane on the main road are not set with separation facilities.
④ Illegal motor vehicle parking occupies bicycle lane.

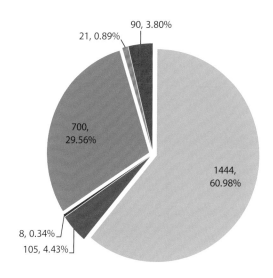

调查：宝华路南段断面交通流量
Research: The traffic flow in the south section of Baohua Road

缺乏物理隔离和严格监管，无法阻止机动车占自行车道停放。
Lacking of physical separation and supervision cause motor vehicle's occupancy on bicycle lane.

Chapter Three

狭窄的自行车道被货物和违章停车占用。
Narrow bicycle lane is occupied by goods and illegal parking.

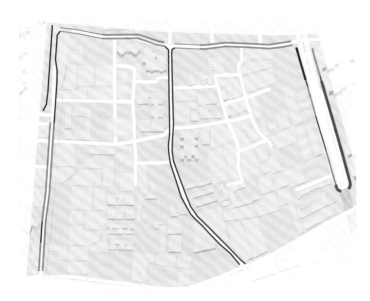

现状自行车道宽度（调研区域）
Current width of bicycle lane (researched area)

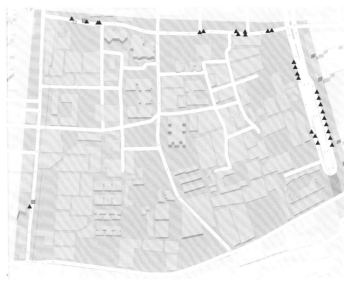

现状自行车道障碍物（调研区域）
Current bicycle lane barrier (researched area)

改善措施 Improvement Measure

①流量较少、车速较慢的车道可不设置自行车专用道，但需实施交通宁静措施；
②在单行道路段可为逆行的自行车流设置专用道；
③车速高、流量大的主干道上（如康王路、中山八路）应设置连续的、有物理隔离的自行车道，隔离方法多样，可因地制宜；自行车道宽度不小于 2 米；
④加强停车管理，设置物理隔离也能有效防止机动车占用自行车道。

① The bicycle lane is not necessary for the motor vehicle lanes with less traffic flow and slow speed, but the implementation of traffic noise-off measures is needed.
② Bicycle lane can be set up in the one-way road for the bicycles flow in different direction.
③ The main road with high speed and large flow （Kangwang Road, Zhongshanba Road） should be set up with consistent bicycle lane with physical separation. But for the bicycle lane should be constructed in consist with different road condition. The width of bicycle lane should not less than 2 meters.
④ Strengthen the parking management; setting physical separation is also an effective way to prevent motor vehicles' occupancy on bicycle lane.

利用路边停车、自行车停放点及公共自行车隔离并保护自行车道，结合隔离带设计。（巴塞罗那）

Use roadside parking, bicycle parking lot and public bicycle to isolate and protect its own road, combining with the design of the isolation belt. (Barcelona)

Chapter Three

在共享的单行路段设置逆行自行车专用道。（巴黎）
Set a dedicated lane for the bicycles in shared one- way vehicles' lane. (Paris)

设置自行车专用道有以下几种方式：
①抬升自行车道地面；
②设置护柱、隔离带等物理隔离；
③利用路边停车、自行车停车设施或公共自行车隔离并保护自行车道（需考虑开门距离）；
④为自行车道涂装颜色，在机动车流量较少，允许自行车与机动车共享的单行路段，可设置逆行自行车专用道。

There are several approaches setting bicycle lanes as follow:
① Uplifted bicycle lane.
② Set bollards or physical separation.
③ Adopt roadside parking, bicycle parking separation facilities or public bicycle separation to prevent bicycle lane. (The distance of the open door should be taken into consideration).
④ Paint color on bicycle lane. Bicycles can share with motor vehicles in the same road section when the motor vehicles flow is small. Dedicated bicycle lane can be set for the bicycle in contract direction flow.

专题2：自行车过街设施
Topic 2: Bicycle Crossing Facilities

现状问题 Current Situation

①主干道交叉口没有设置专用、安全的自行车过街通道；
②自行车道在交叉口处断开；
③大型交叉口处没有设置自行车过街等候区。

① Safe bicycle crossing passage is not set in the intersections on main road.
② Bicycle lane's interruption at the intersection
③ Waiting area for bicycle crossing is not set at large intersections.

自行车过街设施设置要点
Key Points of Bike Crossing Facilities

①自行车在过交叉口时，应有其专用过街通道，与行人分离，各行其道，以提高安全性。与行人过街相似，自行车过街也应需安全岛保护。
②交叉口处的自行车道应重点突出，提醒机动车司机注意通过的自行车。
③应在机动车停车线前设置自行车等候区，在绿灯放行时，自行车可优先启动通过交叉口，减少与机动车的冲突。

① When bicycles cross the intersection, they should have their special crossing passage to separate them from the pedestrians. Thus, enhance the safety. Similar to pedestrians, bicycle also need safety island.
② The bicycle lane at intersections should be highlighted to remind motor vehicle drivers to aware of the crossing bicycles.
③ Bicycle crossing waiting area should be set in front of the vehicle stopping line. When the light turns to green, bicycles can start prior to the vehicles, which reduces the conflict between bikes and vehicles.

交叉口处，自行车等候区位于机动车停车线前方。右转及直行的自行车在绿灯时优先启动，不受直行和右转的机动车阻碍，同时，左转的自行车也不受直行车辆的干扰。（日内瓦）

At the intersection, the bicycle waiting area should be located in front of the motor vehicle parking. The bicycles which turn right or go straight can start firstly when the green light released. At the same time, the bicycles which turn left won't be interfered by the vehicles which go straight. (Geneva)

Chapter Three

自行车与行人过街分离，自行车过街也需要安全岛保护，道路空间充足、车流量大时，需区分不同方向的空间。（乌特勒支）

Bicycle and pedestrian crossing should be separated, and bicycle crossing should also be equipped with safety island. When space is sufficient and the traffic is large, it is necessary to distinguish different directions. (Utrecht)

设置自行车专用过街通道，并用红色强调自行车道，加设自行车信号灯。（阿姆斯特丹）

Set bicycle crossing, highlighted in red and added a signal lamp. (Amsterdam)

改善措施 Improvement Measure

①在主干道交叉口设置专用的自行车过街通道；并设置安全岛作为保护；
②在交叉口处，使用颜色区分自行车道，引起机动车司机的注意；
③大型交叉口处设置自行车过街等候区，确保自行车优先通行。

① Set bicycle crossing passage at the intersection on main road, and set safety island for protection.
② Use red bicycle lane at the intersection to remind motor vehicle drivers.
③ Set bicycle crossing waiting area at large intersection to ensure the bicycle traffic priority.

改造方案：Transformation Scheme

①为交叉口处的自行车道涂装醒目的颜色，引起机动车司机的注意，提高骑车者过街的安全性。
②增加自行车过街等候区，并以彩色涂料示意，供左转和直行的自行车等候使用。
③拐角处的道路空间对机动车毫无作用，可供行人使用。斑马线可以向交叉口移动，使过街的距离更短，行人更安全。这些设计都不会影响机动车的正常通行。

① The bicycle lane at the intersection should be painted in red to remind motor vehicle drivers and raised the safety for the crossing bicycles.
② Add bicycle waiting area and have it painted for the bicycles which turn right or go straight.
③ The corner of the vehicle space has no actual function, which can be used for pedestrians. Zebra crossing can be moved towards the center of the intersection to shorten the crossing distance, so that pedestrians are safer. Meanwhile, the vehicles would not be affected.

交叉口现状 Intersection Status

改造方案示意图 Proposal

Chapter Three

专题 3：自行车停放设施

Topic 3: Bike Parking Facilities

现状问题 Current Situation

① 自行车停放需求大，现状缺乏自行车锁桩及保管处，且分布不合理；
② 已有自行车锁桩只能锁住一个车轮，不够安全；

① The demand for bicycle parking is huge. However, the current bicycle parking environment is lack of locks and custody departments, and their coverage is not rational.
② The existing locks can only lock one wheel of the bicycle. The bicycle parking environment is still lack of safety.

许多地方不可以停放自行车，有的甚至把自行车伸到机动车道上，非常不安全。

Many places can't lock bicycles, and some even lock their bicycles on motor vehicles, which is totally unsafe.

大量居民骑车出行，人们利用骑楼柱子及树木锁自行车。这样充分利用了行道树之间的多余空间。

A large number of residents travel by bicycles. People use arcade pillars and trees to lock their bicycles. It makes full use of the extra space between the trees.

实践案例
Practice

仅有的自行车锁桩只能锁前轮，不够安全，而且位置有限。

The existing bicycle locking facilities can only lock the front wheel. It is not safe and the number of the locks is limited.

街巷内有少数几个自行车保管站。

There are only few bicycle custody departments in the street.

自由停放的自行车量 The number of bicycles which are freely parked	自行车保管站停放总数 Total number of bicycles which parks in custody departments	需求总量 Total demand	自行车保管站容量（供给总量） The volume of bicycle storage capacity in custody departments
104	375	479	470

非正式自行车停放及自行车保管站（调研区域）

Informal bicycle parking and bicycle custody departments (Researched area)

Chapter Three

良好的自行车停放设施
Good Bicycle Parking Facilities

为个人自行车的使用者提供安全、便捷的停放设施，能鼓励更多的人使用自行车出行。自行车停放设施的设计应考虑与周边建筑环境的协调、当地的气候环境（封闭和开敞式的选择）、防盗系统及节省空间的停放方式。

Providing safer and more convenient parking places to bicycle users can encourage more bicycle travel. The design of parking places should be construced in consists of the surrounding environment (close or open) and building, anti-theft system and the measure of saving space.

停车盒子可挡雨，更安全。（阿姆斯特丹）
A parking box should be provided as the sheltering and safer parking place. (Amsterdam)

一个机动车位能停放的自行车数量可达10辆以上，在空间有限的区域，用一个机动车停车位作为自行车停放点是很好的解决方法。（纽约）
A motor vehicles parking lot can park more than 10 bicycles. In the space of limited areas, use one parking space as the bicycle parking place is a very good solution. (New York)

外拓人行道设置"U"型停车架，这种停车架不但能锁上自行车前后轮或车架，停车架可两侧停放，存取车方便安全，而且能提高了停车设施的效率。（悉尼）
U-shape bike racks are installed on the sidewalk extension. These racks can lock both the front and the rear wheels of bikes, which is easy and safe. These are space-saving, as bikes can be put on both sides of the racks. (Sydney)

实践案例
Practice

室内停车架

Indoor parking stand

改善措施 Improvement Measure

①在空间有限的人行道上，可设置"U"型锁桩，可锁上前后两个车轮，比较安全，也节省空间；
②自行车停车需求大，竖向空间较充足的地方，可考虑采用双层自行车架，能更有效地利用空间满足大量的停车需求；
③可把一个路边机动车停车位变为自行车停放点；
④在地铁站周边及建筑内部设置自行车停放区。
⑤自行车停放点的位置应根据需求设置。

① In the places of limited space one can set up "U" type locks to lock two wheels.It is safer and saves more space.
② In the place with large bike parking demands and sufficient vertical space, double-layer bike racks can accommodate the demand and save space.
③ Transform the side parking spaces into bicycle parking areas.
④ Set up bicycle parking area near the subway station and inside the building.
⑤ The location of bicycle parking should be set according to the different demand.

充分利用天桥桥底空间，更能为自行车架遮阳挡雨。（广州）

Making full use of the space under the bridge can also use as the shelter for bicycle. (Guangzhou)

社区外自行车停放点，双层自行车架有效节省空间。（东京）

The parking place outside street block, Double layer bicycle parking frame can effectively save more space. (Tokyo)

Chapter Three

1.3 交叉口
Intersection

现状问题 Current Situation

①大部分交叉口都有斑马线及信号灯，但无障碍通道不完备。
②较大的交叉口缺乏行人安全岛，行人过街不安全、不方便。
③机动车转弯半径过大，且没有渠化岛，车速过快，行人过街不安全。
④交通信号配时不合理，行人、自行车等待时间过长。
⑤较大的交叉口没有设置自行车专用过街通道。
⑥交叉口没有导向标识。

① Most of the intersections have zebra crossing and signal lights, but the barrier free passage is not completed.
② The wide intersections are lack of pedestrian island, which makes pedestrian unsafe and inconvenient.
③ Motor vehicle turning radius is too large, and there is no channelized island. The speed of motor vehicles is too fast. It is unsafe for pedestrians.
④ The time of the traffic signal is unreasonable. The time for pedestrians, bicycles' to wait is too long.
⑤ The wide intersection is not set with dedicated bicycle lane.
⑥ Intersection has no orientation identification.

宝华路与多宝路路口面积大、人流密集，但机动车流量少、方向少，应减少宝华路的车道数量，增加渠化岛，使其成为行人优先区域。

The intersections in Baohua Road and Duobao Road are large and crowded, but the vehicle flow is small. The number of Baohua road lanes should be reduced for increasing the channelization island to make it become the pedestrian priority area.

实践案例
Practice

斑马线端头没有设置无障碍通道。
The zebra crossing is not provided with a barrier free passage.

无障碍通道设计不合理。
The design of barrier free passage is unreasonable.

无障碍通道被人为地变成"有障碍"。
The barrier free passage is artificially transformed into a "barrier"

097

Chapter Three

改善措施 Improvement Measure

① 确保所有斑马线两端设有无障碍通道，设计规范合理的坡道。
② 减少机动车不必使用的空间；适当减少转弯半径，在交叉口较大的地方增加渠化岛，降低车速；扩充部分人行道，把行人和自行车过街位置往交叉口移，减短过街长度。
③ 增加或重新合理分配交叉口信号相位。
④ 在较大的交叉口设置自行车专用过街通道。
⑤ 把人流较大的交叉口抬升，作为步行优先区域，使机动车减速通行。
⑥ 在交叉口加装行人导向标识。

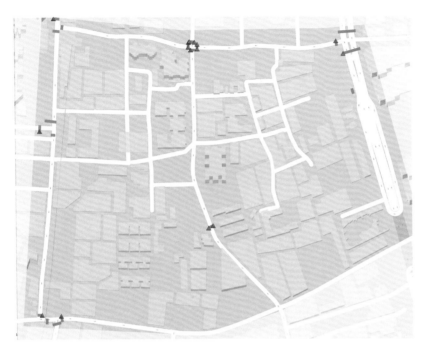

调研区域过街设施调查
The crossing facilities survey in researched area

图例 Legend
▲ 无残疾人坡道处 Place lack of barrier free ramp
— 行人过街 Pedestrian crossing

① To ensure that all zebra crossings are provided with barrier free access and the design of the ramps are reasonable and specified.
② Limiting the space where motor vehicles won't use, appropriately reduce the turning radius. Increase channelized island at the area of large intersection thus reach the purpose of reducing the speed of motor vehicles. Move the bicycle and pedestrian crossing to the intersection for shortening the length of street crossing.
③ Increase the number or reallocate the intersection signal.
④ Set bicycle lane at the places with large intersection.
⑤ Raise the intersection with large pedestrian volume to make it a pedestrian-priorized area, slowing down the traffic speed.
⑥ Install the pedestrian guiding signs at intersections.

专题1：扩大行人空间，减少不必要的机动车空间

Topic 1: Expanding Pedestrian Space, Reduce Unnecessary Vehicle Space

如果设有路边停车带，应把交叉口的人行道扩大到路边停车带的外侧边线，缩短过街距离。（里昂）

If there is a paking zone at the roadside, the sidewalk at the intersection should be expanded to the outer edge of the parking zone to shorten the distance between the crossing. (Lyon)

在项目改造前期，可以采用临时措施，比如使用护柱和划定禁行区，在试验了实施效果之后，就可扩大人行道的铺装。（名古屋）

In the early stage of the project, temporary measures can be provided if necessary, such as the use of bollard and designation of restricted area. Sidewalk can be expanded after the implementation of the test. (Nagoya)

通过人行道向车行道扩张，增加慢行交通系统空间，减少机动车流量，同时也减少行人、非机动车暴露在机动车流中的时间。

Expanding the sidewalk to the outer area can increase more space for non-motor traffic system, at the same time reduce the time of pedestrian and non-motor vehicles exposuring in the motor vehicles' flow.

Chapter Three

改造方案 Transformation Scheme

该交叉口有两个拐角（左上及右下）是不允许车辆转弯的，因此不需要这么大的转弯半径。而且由于现状人行道非常窄，转角的位置有很多电线杆、标识牌，行人通行和等候过街的空间不足，所以可拓展人行道的空间。在近期实施时，可先用护柱限制机动车通行的空间。

The intersection has two corners that are not allowed to turn around, so it is not required for large turning radius. And present condition of the sidewalk is very narrow. There are a lot of wire or signages or others at the corner. It lacks the space for pedestrian to cross the street. Therefore, the space of sidewalks can be expanded. Bollards can be adopted to limit the space of motor vehicles in the near future.

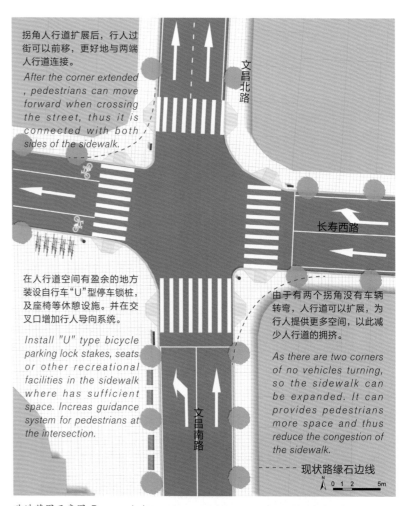

拐角人行道扩展后，行人过街可以前移，更好地与两端人行道连接。
After the corner extended, pedestrians can move forward when crossing the street, thus it is connected with both sides of the sidewalk.

在人行道空间有盈余的地方装设自行车"U"型停车锁桩，及座椅等休憩设施。并在交叉口增加行人导向系统。
Install "U" type bicycle parking lock stakes, seats or other recreational facilities in the sidewalk where has sufficient space. Increas guidance system for pedestrians at the intersection.

由于有两个拐角没有车辆转弯，人行道可以扩展，为行人提供更多空间，以此减少人行道的拥挤。
As there are two corners of no vehicles turning, so the sidewalk can be expanded. It can provides pedestrians more space and thus reduce the congestion of the sidewalk.

改造范围示意图 Proposed plan

实践案例
Practice

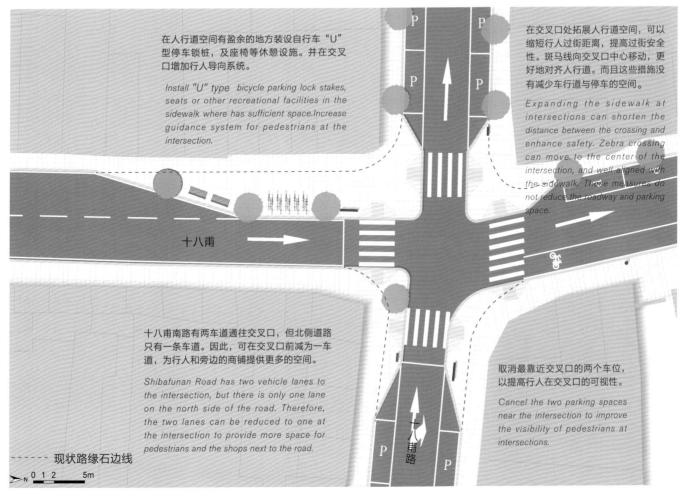

在人行道空间有盈余的地方装设自行车"U"型停车锁桩，及座椅等休憩设施。并在交叉口增加行人导向系统。

Install "U" type bicycle parking lock stakes, seats or other recreational facilities in the sidewalk where has sufficient space. Increase guidance system for pedestrians at the intersection.

在交叉口处拓展人行道空间，可以缩短行人过街距离，提高过街安全性。斑马线向交叉口中心移动，更好地对齐人行道。而且这些措施没有减少车行道与停车的空间。

Expanding the sidewalk at intersections can shorten the distance between the crossing and enhance safety. Zebra crossing can move to the center of the intersection, and well aligned with the sidewalk. These measures do not reduce the roadway and parking space.

十八甫南路有两车道通往交叉口，但北侧道路只有一条车道。因此，可在交叉口前减为一车道，为行人和旁边的商铺提供更多的空间。

Shibafunan Road has two vehicle lanes to the intersection, but there is only one lane on the north side of the road. Therefore, the two lanes can be reduced to one at the intersection to provide more space for pedestrians and the shops next to the road.

取消最靠近交叉口的两个车位，以提高行人在交叉口的可视性。

Cancel the two parking spaces near the intersection to improve the visibility of pedestrians at intersections.

- - - 现状路缘石边线

改造方案平面示意图 *Proposed plan*

101

Chapter Three

专题 2：抬升交叉口

Topic 2: Raised Intersection

抬升交叉口地面，使用红色铺装，加装护柱限定机动车行驶范围。（布达佩斯）

Raise intersection, pave red pavement and install bollards to limit the vehicle access. (Budapest)

抬升整个交叉口地面。（奈梅亨）

Raise the entire intersection. (Nijmegen)

改造方案 Transformation Scheme

现状交叉路口已经改造成行人铺装，但交通宁静效果不够明显。可抬升整个交叉口地面，并重新组织德兴路及杨巷路的车道，在德兴路增加自行车道，并以路边停车对其隔离及保护（需考虑开门的距离）。

In the current situation, the intersection has been transformed into a pedestrian pavement, but the effect is not obvious. The entire intersection can be uplifted at the same time reorganize the lanes in Dexing Road and Yangxiang Road. Increase bicycle lane at Dexing Road, and use separation and protection for the bicycles that park along the roadside.

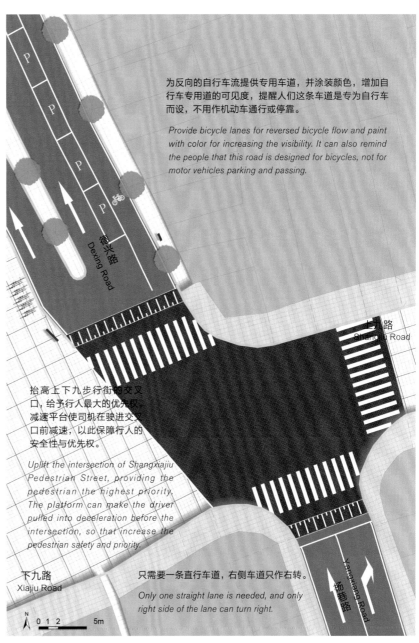

为反向的自行车流提供专用车道，并涂装颜色，增加自行车专用道的可见度，提醒人们这条车道是专为自行车而设，不用作机动车通行或停靠。

Provide bicycle lanes for reversed bicycle flow and paint with color for increasing the visibility. It can also remind the people that this road is designed for bicycles, not for motor vehicles parking and passing.

抬高上下九步行街的交叉口，给予行人最大的优先权，减速平台使司机在驶进交叉口前减速，以此保障行人的安全性与优先权。

Uplift the intersection of Shangxiajiu Pedestrian Street, providing the pedestrian the highest priority. The platform can make the driver pulled into deceleration before the intersection, so that increase the pedestrian safety and priority.

只需要一条直行车道，右侧车道只作右转。

Only one straight lane is needed, and only right side of the lane can turn right.

改造范围示意图

Proposed plan

Chapter Three

专题 3：信号灯相位

Topic 3: Signal Lamp Phase

影响行人过街行为的除了过街的位置和种类，还包括信号灯相位，即行人等待时间。调查发现，如果行人等候过街超过 30 秒，他们将做出冒险的行为，譬如闯红灯跑过马路、离开人行道到马路边缘或者在汽车间穿行等。根据各国经验，行人信号灯等候时间不应该超过 60 秒。

In addition of location and type of the intersection, the elements which affect pedestrians crossing include signal lamp phase, the time for pedestrian's waiting. Survey found that if pedestrians wait the signal light t for more than 30 seconds, they will make risky behavior. For example, run to go through red light, or leave the sidewalk to the edge of the road and walk into the motor vehicle flow. According to the experience of many countries, the time for pedestrian's waiting should not exceed 60 seconds.

在荔湾旧城内调查了交叉口信号灯相位，行人等候时间都不超过 60 秒，但没有实施"行人优先通行（LPT, Leading Pesestrian Interval）"相位。

We did an investigation to the intersection in the old city in Liwan. The time for pedestrian's waiting is no more than 60 seconds, but no "LPI" phase is implemented.

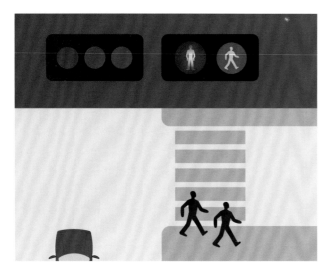

行人与右转车辆禁行。

Pedestrian and right turn vehicles are prohibited

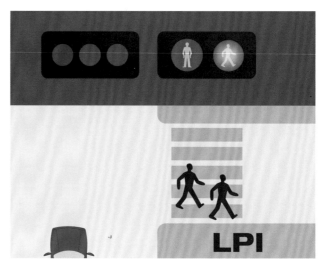

LPI 开始，行人先放行，右转车禁行。

When the LPI begins, pedestrians pass first and vehicles' right turn are prohibited.

行人优先通行 Leading Pedestrian Interval

行人优先通行（LPI, Leading Pedestrian Interval），是指信号灯同一相位放行的行人相位与机动车相位相差数秒，行人比机动车优先放行。同时也有自行车优先通行（LBI, Leading Bicycle Interval）。行人绿灯亮时，同一相位放行的机动车信号灯仍为红色，而 3~4 秒后，机动车灯变绿；这时，行人和自行车已经出现在道路中或通过马路的一半，而右转的车辆才准备启动。这给予行人、自行车过街更充足的时间，使他们比转弯车辆提早通过，提前出现在机动车前，增加机动车减速的可能性。

Leading Pedestrian Interval (LPI) refers to the second difference when the release of green light for pedestrians while the light turns red for vehicles to stop. Pedestrians should have the priority of crossing. At the same time, there is also Leading Bicycle Interval (LBI). When the pedestrian green light is on, the motor vehicle signal lamp is still red for 3 to 4 seconds before it turns green. At this time, pedestrians and bicycles have across half of the road while the vehicles turn right has just started. This provides pedestrians and bicycles enough time to cross the street before the passing of vehicles. It increases possibility of motor vehicles' deceleration when they occur in advance of the motor vehicle.

一项纽约的交通调查显示，实施 LPI 的交叉口，行人交通事故的伤亡率比其他交叉口低 26%，且 36% 的受伤者伤势较轻。

New York transportation survey shows that after the implementation of LPI on the intersection, the pedestrian traffic accident rate is 26% lower than the other intersections, while 36% of the wounded is minor injuries.

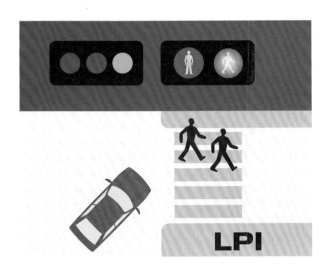

数秒后，LPI 结束，行人继续通行，右转车放行，行人已在斑马线中，增加机动车减速的可能性。

A Few seconds after the end of the LPI, pedestrians continues passing and the vehicles begin to turn right. While pedestrians are already in the zebra crossing the possibility of reducing the speed of the vehicle increase.

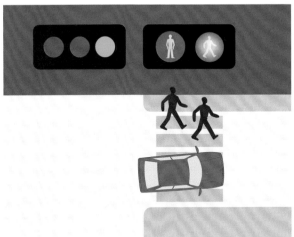

右转到达斑马线时，行人已通过马路的一半，减少与右转车的冲突。

While motor-vehicles reach the zebra crossing, pedestrians have already crossed half of the road. It can reduce the conflict between pedestrians and the vehicles which turn right.

Chapter Three

1.4 路中过街
Mid-Block Crossing

华林玉器街入口

The entrance of Hualin Jade Market

长寿西路华林玉器街入口过街需求很大。

The crossing demand at the entrance of Hualin Jade Market in Changshouxi Road is large.

调查位置的南北两侧都集中了玉器商贩，是广州最集中的玉器市场。因此许多人需要横跨长寿西路。调查区东侧的斑马线是康王路上主要的交叉口之一，人流非常大。但在华林玉器街对出的路面，集中了大量的过街人流，总量超过了斑马线上的过街人流。

North and south sides of the market are full of jade traders. It is the most concentrated jade market in Guangzou. Therefore, many people need to cross Changshouxi Road. The east side of zebra crossing is one of the biggest intersections in Kangwang Road. The population flow is huge. But the opposite side of the Hualin Jade Market concentrateds a large number of people across the street. The total number of people has exceeded the number of people on the zebra crossing.

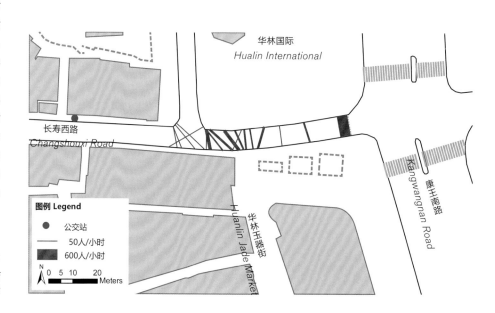

长寿西路玉器市场外行人过街轨迹调查

A survey of pedestrian crossing out of Changshouxi Road jade market

实践案例
Practice

现状问题 Current Situation

调查区域西侧为恒宝华庭和长寿路地铁站出入口，东侧为大笪地购物街和地铁出入口，两侧人流来往非常大。如图所示，在斑马线过街的行人并不多，更多的行人是在宝华路上随意穿行。这条路的车流量不大，但路面却很宽。行人过街最集中的是宝华小区外的路面。宝华中约内是繁忙的市场，人流较大，直接穿越宝华路正好到达恒宝华庭的正门台阶。因此，即使距离斑马线不到10米，人们也不愿绕行到斑马线上。另一个集中的过街点是公交车站。许多乘客下车后就直接横跨马路，而车站北侧50米有红绿灯的斑马线，却较少人使用。应该指出，这些现象并不代表行人的行为不文明，而是道路设计没有切实考虑使用者的需求，没有在过街需求大的地方（如公交车站）提供安全措施。

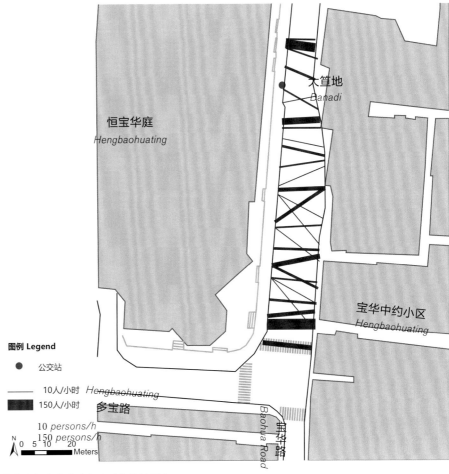

宝华路恒宝华庭外行人过街轨迹调查
Pedestrian crossing trajectory survey in Baohua Road and outside Hengbaohuating

The west side of the researched area is the Hengbaohuating tribunal and Changshou Road subway station entrances. The east side is Dadadi shopping street and the subway entrance, and the population flow is large on both sides of the road. As showed in the image, not many pedestrians crossing street through zebra crossing, instead, more pedestrians randomly across the street on Baohua Road. The traffic flow is not large, but the road is very wide. The most concentrated pedestrian crossing is the road outside of Baohuazhongyue. It is a very busy market with large crowd. Straightly across Baohua road one can reach the front gate stages of Hengbaohuating. Therefore, even if the distance of the zebra crossing is less than 10 meters, people do not want to detour to the zebra crossing. Another concentrated pedestrian crossing point is bus station. Many passengers get off and directly across the road. However, the traffic lights are only about 50 meters north of the station, but fewer people use it. It should be pointed out that these phenomena do not represent the uncivilized behavior of pedestrians, but shows that the road design does not truly consider the demand of users. Security measures are not provided in the street with high demand for pedestrian crossing(such as bus station).

Chapter Three

1) 缺乏路中过街设施

The lack of street crossing facilities in the middle of the road

在旧城，少有路中过街设施。而在调研区域，文昌路（长寿路至下九路段）全长 513 米，只有一处路中过街，平均过街间距为 260 米。根据行人过街习惯，一般在市区内，过街设施之间距离应为 80~100 米，具体根据道路两侧的用地性质决定。

There are few road crossing facilities in the old city. However, in the researched area, the total length of Wenchang Road (Changshou Road to Shangxiajiu Road) is 513 meters. There is only one road crossing, making the average crossing distance 260 meters. According to the pedestrian habits, general in the urban area, the street distance between the crossing facilities should be 80 to 100 meters. Some places needs to be constructed according to the condition on both sides of the road.

宝华路上的行人都习惯随意穿行。
Pedestrians on Baohua Road are used to across the street randomly

大部分路段较窄、车流量较少，即使没有足够的路中行人过街，人们也常随意穿行。而车辆在交通量较少的路段会加快车速，如果没有减速设施或行人过街作提示，就会存在安全隐患。

Most of the roads are narrow with less traffic, people still randomly across the street even there is not enough space in the middle. But vehicles on the road with less traffic, they will accelerate their speed. Unsafe factors remain if there is no any deceleration facility or pedestrian crossing signal.

而在黄沙大道、康王路等多条主干道上，人们只能使用行人过街设施。而这几条道路上的过街间距较长，成为区域的行人交通障碍。

On the main roads like Huangsha Avenue and Kangwang Road, people can only use pedestrian crossing facilities, but the distance between crossing is long. It causes obstacles for the pedestrians in this area.

2) 在过街需求大的地方没有行人过街设施

No crosswalk at the place with high demand of crossing

在旧城街巷内的商业活动非常频繁，这些街巷通往道路的出入口，往往是行人过街需求最大的地方，且这些位置又不常在道路的交叉口处。所以便会出现路中过街的人流比交叉口斑马线过街的人流大的情况。

The commercial activities are very busy on the alleys in the old city. The entrances of these alleys often have high demand of crossing, but these positions are not always at the intersections, so that there are more people crossing on the other sections of the road than those on the crosswalk.

3) 路中过街没有信号灯或减速设施

Crossing without signal lights or deceleration facilities

即使在部分路段设有路中行人过设施，如文昌路，但并没有装设信号灯或减速设施，无法确保行人过街的安全性。

Even there are some road sections with pedestrian crossing, such as Wenchang Road, but it is not equipped with signal lamp or deceleration facilities. It cannot ensure the safety of pedestrians to cross the street.

行人往来宝华小区都不走在斑马线上。

Pedestrian in Baohua Road do not walk on the zebra crossing.

Chapter Three

改善措施 Improvement Measure

1) 增加路中过街设施

Add mid-block crossing

在路段较长或主干道上增加路中过街设施，缩短过街距离，提高步行的便捷性。尤其是要增加黄沙大道、康王路、中山七路及中山八路的行人过街设施，有助于加强西郊游泳场、黄沙海鲜市场、沙面以及陈家祠等重要公共设施和景点与旧城内部街区的连接。

Install road crossing facilities on the main road or the road section with long distance to shorten the space and improve the convenience of walking. Particularly increase the number of road crossing facilities in Huasha Avenue, Kangwang Road, Zhongshanqi Road and Zhongshanba Road. It will help to strength the connection between the public facilities like Xijiao Swimming Pool, Huangsha Seafood Market, Shameen Chen Clan Acaderny and other urban internal blocks.

2) 行人过街设施的位置应根据过街需求而定

The location of pedestrian crossing should be based on the requirement of the demands

应充分了解区域内行人的出行习惯，在人流大的重点区域和过街需求大的地方设置行人过街设施。

Install facilities after fully understanding of the pedestrian travel habits in the region, the flow of people in the key areas, and places with large demand for road crossing.

3) 为过街设施配备信号灯或减速设施以确保行人的安全性

Equip with street lights or deceleration facilities in pedestrian crossing, to ensure the safety of pedestrians

不需要为所有的行人过街设施安设信号灯，但在没有信号灯的过街处，需要增加减速措施。其中一个可行有效的措施是抬升过街地面，并涂装成显眼的红色。把斑马线抬升至与人行道同高，这样轮椅或婴儿车将不受高差的影响，而且车辆驶近时会因为坡度而自觉减速。将过街区域涂装成红色，也可提醒司机减速。如果在重要的街巷外，如玉器街、西关博物馆等入口处，还能起到游人导向功能。

Do not need to install signal lights at all the pedestrian crossing, but should set deceleration facilities instead. One of the feasible and effective measures is to uplift pedestrian crossing and paint it in red color. Uplifting the zebra crossing to the same height with sidewalk, so wheel carts or baby carriages will not be affected by the height difference. And the vehicles will slow down their speed when they are approaching. Change the crossing area into red color can remind drivers to slow down. If the red painting is in the streets, such as Jade Street, Xiguan Museum entrance, it can also be seen as an attraction or guidance for visitors.

路中过街设计要点 Design Key Points of Crossing

①根据行人过街需求设置过街的位置，人流大的地方应设置较宽的斑马线。
②抬升行人过街。行人过街抬升至与人行道同高，方便无障碍通行。同时上下坡处应划有长短线，地面涂装红色以提醒驾驶者。在人行道边缘安装护柱，防止机动车驶入人行道。
③设置减速带或其他交通宁静设施。可在斑马线前几米设置减速带，确保司机有足够的刹车距离。
④在较宽的马路中设置安全岛。过街距离超过两车道时，应该设置安全岛，安全岛宽度需保证能容纳轮椅、婴儿车或自行车。
⑤增加限速等其他标识。
⑥根据道路的宽度、车流及车速限制选取适当的过街设施。

① The location setting of pedestrian crossing should according to the demand of road crossing. Places with large population flow should be equipped with wider zebra crossing.
② Uplifting pedestrian crossing. Pedestrian crossing should be uplifted to the same height of sidewalk. There should be long and short lines on the upper and down ramps. The ground should be painted with red color to remind the drivers. Install bollard at the edge of sidewalk to prevent motor vehicles entering into sidewalk.
③ Set deceleration zone or other facilities to deaden the noise of traffic. Speed bump can be set few meters in front of the zebra crossing to ensure that the driver has enough distance for braking.
④ Set up safety island in wide road. The street crossing which exceeds two motor vehicle lanes should be equipped with safety island to ensure that it is safe for wheelchairs, baby carriages or bicycles.
⑤ Add speed limit and other signals.
⑥ Select the proper crossing facilities according to the width of the road, traffic flow and speed limit.

确保简单、直接的平面过街。（东京）

Ensure simple, direct street crossings always at street level. (Tokyo)

连续的抬升的人行道能给步行者优先权。（布里斯班）

Continuous sidewalks over streets gives pedestrians priority. (Brisbane)

Chapter Three

无论有无信号灯,都应在较宽的马路上设置行人安全岛。(荷兰)

Pedestrian safety island should be equipped on the wide road, no matter the existence of signal light.(the Netherlands)

在斑马线前数米设置减速带,路中设置安全岛和灯光指示。(香港)

Set deceleration bump few meters before the zebra crossing, and set up safety island and guidance in the middle of the road. (Hong Kong)

行人过街和自行车过街抬升,并把人行道往外扩,减少过街距离。(巴黎)

Raise the crosswalk and bike crossing, and extend the sidewalk to reduce crossing distance.

改造方案 Transformation Scheme

龙津西路是市民和游客到达荔枝湾涌的重要路径，这里还是典型的骑楼街，并连接着西关大屋保护区和西关博物馆，是旧城最重要的景观街道之一。虽然这里的车流不大，但行人穿越马路的安全性应该得到进一步提升。

Longjinxi Road is an important path to Lizhiwanchong for both the public and tourists. There are traditional arcade street, which connects Xiguan house protection zone and the Xiguan Museum. It is one of the old city's the most important streetscapes. Although the traffic flow here is not large, the safety of pedestrians crossing should be further enhanced.

建议在龙津西路荔湾博物馆牌坊外设置抬升的行人过街，一方面方便市民安全通行，另一方面整理逢源西二巷巷口与人行道的高差。而且，现在的西关博物馆入口牌坊较不够显眼，入口也较小，红色的抬升过街还可起到游览导向作用，吸引游人的注意。

Recommend to set raised crosswalk outside the arch of Liwan Museum in Longjinxi Road. On one hand, it makes citizen's travel safe and convenient. On the other hand, it unifies the height between the sidewalk and the access of Fengyuanxi Second Alley. The arch of Xiguan Museum entrance is not conspicuous, and the entrance itself is also small. The raised crosswalk in red can guide and attract tourists.

行人过街抬升，上下坡处划有长短线提醒司机，在与人行道接驳处细心考虑无障碍通行，并配有限速标识。（阿姆斯特丹）

Uplifted pedestrian crossing; the short and long lines on the ramp is for reminding the drivers, and it perfectly connects the sidewalk, which shows a consideration of barrier free passing, and is also equipped with deceleration signal. (Amsterdam)

现状
Current status

改造方案示意图
The scheme diagram of reconstruction

Chapter Three

1.5 行人导向系统
Pedestrian Guidance System

现状问题 Current Situation

①路径关键节点缺乏导向标识。旧城区现有的导向标识不多，大部分位于荔枝湾涌周边。在道路交汇及转折处，或者地铁站出入口极少发现导向标识。
②现有导向标识高度太低，在人流拥挤的地方不容易被发现。而且整体颜色与环境背景过于相似，无法引起行人的注意。
③指示牌包含的信息过于单一，没有地图，只有名字与箭头，不足以清楚引导行人，尤其是对于不熟悉情况的旅客，只有地名是不足够的。
④字体与背板的颜色对比度较低，难以阅读。背板的木色与字体的绿色对比太弱，尤其是当木色掉落后，字体更难以识别。
⑤缺乏系统性的、统一的设计。现在旧城使用的标识系统大概有3个，一个是全市通用的棕色的旅游交通标识，一个是"上下九"的标识系统，第三个是荔枝湾涌周边使用的木牌。导向系统过多，信息没有整合，不能让游客迅速掌握导向信息。

① The Lack of guiding signs at the key junction of the road; there are not many signs in the old city, and most of them are set around LizhiwanChong. At the intersection of the road and the turning point, or the entrance of subway can scarcely find guiding signs.
② The existing signs are too low to be noticed in the crowded places. And the overall color is too similar with the environment background, which cannot attract pedestrians' attention.
③ The information on the signs is too simple, without map on it, only names and arrows. It is not clear enough to guide the pedestrians, especially for travelers who are not familiar with the situation. Only a place name is not enough.

指示牌的高度太低，不容易被发现。
The height of the guiding board is too low to be noticed.

④ The contrast between fonts and background color is weak so that it is difficult to read. The wood color background and green color fonts are in weak contrast, especially when the wood color fades.
⑤ Lack of systematic and uniform design. Now the old city has about three guiding signs systems. One is a general brown tourism traffic signs which are used in the whole city, the other one is a guiding system in Shangxiajiu, and the third one is the wood board used around Lizhiwanchong. The guiding systems are too many, and the information has not been organized. It cannot make visitors quickly grasp the guiding information.

指示牌摆放的位置不够显眼，颜色与周边环境颜色相近，而且字体与背板对比度也不足。

The position of the guiding board is not conspicuous, the color is similar with the surrounding environment, and the contrast between the font and the back board is not enough.

Chapter Three

行人导向系统规划策略 Strategy of Pedestrian Guidance System Planning

① 清楚解释周边环境，提供清晰和必需的信息：
- 行人所在位置
- 周边有哪些重要地点
- 重要的目的地及推荐的线路
- 线路的长短和所需的时间
- 途中经过的其他重要地标
- 安全的行人过街位置
- 公共交通的选择

② 提供可读性强的地图指引
地图信息应该根据行人的读图习惯，提供足够的细节，使用易于读懂的表达方式。

③ 有效推广地方形象、商业和历史文化导向系统设计应体现地方文化，有利于旅客对城市保留深刻的印象，并为旅客指示重要的历史文化景点，为人们设计游览线路。

④ 显示区域周边的环境和对外的公共交通信息。

⑤ 区域内有统一的系统让第一次来访者可以迅速熟悉导向系统的使用方法，并能运用这个经验去识读区域内的其他导向标识。

⑥ 导向系统应设置在关键的地方：
- 公共交通站点（地铁及公交站）
- 道路交叉口及转弯处
- 重要地标及游览路径附近

在道路交叉口及行人过街处依然保持连续性，并根据不同地面铺装改变红线的材质。

keep the continuity at the intersection and the place where pedestrians cross and change the red lines texture according to different pavements.

在景点外、转角处及道路分叉口的地面安装"自由之路"的特有标识。

Install the unique marking of "Freedom Trail" on the ground outside the attractions, at the corners and intersections.

⑦形象突出，容易辨识导向系统的形象应非常突出，使行人在拥挤的环境能快速找到标识牌的位置。面板的颜色及与文字的对比度应该足够强烈，让人们在较远的距离能清晰阅读。

⑧设计形象统一的几种标识牌形式，以适用于不同的空间

⑨信息显示应能让大部分人阅读和理解信息显示的高度应考虑儿童、轮椅使用者及其他成人的视线范围，显示方式应考虑视障人士的需求。

⑩开发其他配套导向指引工具：
· 利用二维码，让行人把地图和旅游信息下载到手机或其他移动设备中使用
· 设计与街道导向系统成套的纸质地图

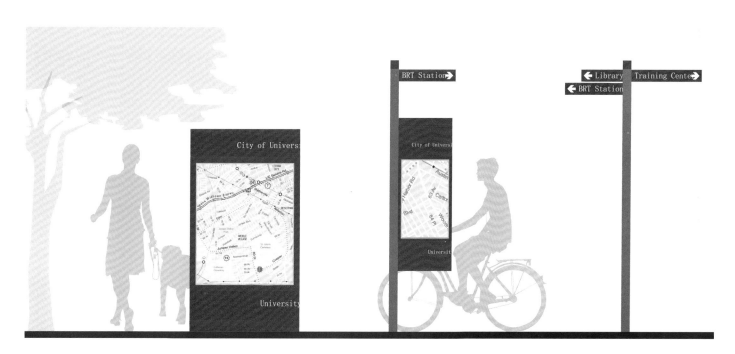

全区域地图
Whole area map

信息指引面板
Information guiding board

单独的方向指示
Direction indicating column

建立统一的导向系统，发展多种标识牌形式，因地制宜使用。
Establish a unified guidance system, a development of several forms of guiding signs to apply to different local conditions.

Chapter Three

① Explain the surrounding environment clearly, and provide clear and necessary information.
- location of pedestrian
- the surrounding important sites
- important destination and recommended line
- the length of the line and the time needed
- other important landmarks along the road
- safe location for pedestrian crossing
- public transport options

② Provide readable map information, which should be based on the habits of pedestrians. Provide enough detail and use easy expression.

③ Effective promotion of local image, business and history and culture. The guiding system design should reflect the local culture, and should be conducive for travelers to keep a deep impression of the city, and also indicates the important cultural and historical sites and make tourist routes for travelers.

④ Present the surrounding environment and public transport information.

⑤ There should be a unified system which provides an easy and quick way to learn about the system for the travelers' first visits. And ensure that travelers can use such experience on other guiding signs within the area.

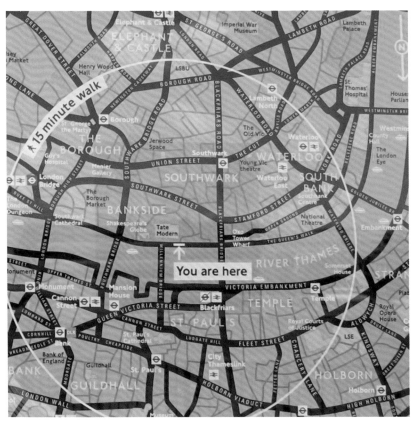

地图显示"您所在的位置",方向与读者面向地图的方向一致,使读者迅速在脑海中建立地图形象,容易辨识自己的前进方向。

Map will show the location of your position, and the direction is the same as in the map which makes reader can instantly create a map image in their mind, and can easily identify their own way forward.

实践案例
Practice

⑥ Set in key areas
· Public transportation stops (subway stations and bus stations)
· intersections and turning point
· important landmarks and nearby tourist routes
⑦ The image of wayfinding system should be prominent and identifiable, ensuring people can easily find it out in the crowded areas. Colors of the background and the texts should have strong contrast, which allows people are able to read in distance.
⑧ The design of the image can be unified in several forms to apply to different space.
⑨ The information should be able to be understood for most of the people
The height of the displayed information should consider the vision range of children, wheelchair users and other people. The presenting mode should also consider the needs of visually impaired people.
⑩ Develop other supporting guidance tools:
· use two-dimensional code to allow people to download maps and travel information on mobile phones or other mobile devices.
· designed the paper map which is compatible with street guiding system

地图上标有建筑外形、公共自行车站点、地铁出入口、出租车站，还有5分钟骑行、5分钟和15分钟步行范围。

Maps show 3D buildings, bike sharing station, metro access, taxi bay, 5 minutes cycling area and 5 minutes and 15 minutes walking areas.

Chapter Three

2 综合改造方案
Comprehensive Proposals

方案：宝华路改造方案

Plan : Comprehensive Proposals on Banohua Road

现状问题 Current Situation

①行人过街流量大，尤其在巷口及公交站附近。交叉口人流量大。行人过街设施不足，而且没有设置在需求最大的地方。
②路段车流量少，路面太宽，车道太多，但与之相连的路段只有1~2条车道。
③东侧人行道过窄。
④交叉口面积过大，转弯半径大，车辆随意穿行，没有减速设施，与大量的过街行人产生极大的矛盾。
⑤路内停车占有率没有达到饱和，有不少长时间停放的车辆。
⑥自行车流量大，但没有专用的自行车道。
⑦缺乏行人休憩的位置。

伦敦牛津街是英国最著名的购物街，人流非常大，在路中设置了一定高度的隔离带，可让行人随意通行，并有安全的等候区域。

Oxford Street in London is the most famous shopping street in the UK with large pedestrian flow. A raised medium provides a safe waiting area, which allows people crossing the street easily at any places.

① The pedestrian flow is large at the crossings, especially those near the alleys and the bus station.
② The traffic flow is less in the road section; the road is wide with too many lanes, but only 1~2 lanes are connected with Baohua Road.
③ The east side of the sidewalk is too narrow.
④ The intersection area is too large with large turning radius. The vehicles passing randomly with no deceleration facilities. It causes big conflict with a large number of pedestrians.
⑤ The roadside parking occupancy rate does not reach saturation, and there are many motor vehicles park on the road for a long time.
⑥ Bicycle traffic is large but no dedicated bike lane.
⑦ Lack of pedestrian rest position.

可通行的路中隔离带种有胸径较小、分支点较高的乔木，不仅美观，而且不阻挡司机发现行人的视线。（伊斯坦布尔）

There are trees with small branches planted in the separating belt, which is beautiful and will not obstruct the drivers' sight on pedestrians. (Istanbul)

改善措施 Improvement Measure

①抬升交叉口及部分路段路面，使其成为行人优先通行区域。
②在公交站后增加路中行人过街设施。
③减少车道数量，只保留两条车道。在交叉口处设置安全岛和护柱，规范机动车行走路线。
④拓宽人行道，增加座椅、公共自行车、自行车停车架及行人导向标识等街道设施。
⑤在路中增加抬升的隔离带，可供行人通行或等候，并提供一些开口，供自行车和手推车过街。
⑥设立有物理隔离的自行车专用道。
⑦取消少量路边停车位。加强停车管理。引导装、卸货车辆到其他专用区域作业。

护柱的可见性非常重要，尤其在交叉口、出入口及转弯的关键位置，必须安装柱体较高、颜色较明亮的护柱。

The visibility of the bollard is very important, especially in the intersection, entrance and key turning places. High bollard in contrast color must be installed in these places.

Chapter Three

① Lift the intersection and some sections of the road, making them pedestrian priority area.
② Add pedestrian crossing behind bus station.
③ Reduce the number of lanes to two lanes. Set up refuge island and bollards at the intersection to regulate the route of motor vehicle.
④ Broaden the sidewalk and increase the number of the seats, public bicycles, bicycle parking and pedestrian guiding signs or other street furniture.
⑤ Add lifted isolation belt in the middle of the road for pedestrians' waiting and passing. Some openings should be provided for the passing of bicycles and carts.
⑥ Install physically separated bicycle lane.
⑦ Cancel a small number of roadside parking spaces and strengthen the parking management. Guide the loading and unloading vehicles to the other special area.

长寿路地铁站 A 出口的广场北侧，可以开辟一段 2 米宽的装卸货区域，货物可由手推车转运至附近的零售商铺。这里设置装卸货港湾，相比宝华路上设置有几点优势：这里为支路，而且是单行道，对交通影响较少，而且不会打断或占用自行车道，现有的出口广场也为其提供了充足的空间。

The north side of subway entrance A in Changshou Road can be explored a 2-meters-wide loading and unloading area. The goods can be transported to nearby retail stores by carts. There are a few advantages to install loading and unloading area here compared to be put in Baohua Road: This is a branch and a one-way street. It is less influenced by the traffic, and will not interrupt or occupy bicycle lanes. The existing exit square also provides sufficient space.

人行道边缘可设置坐高较宽的护柱充当座椅，供行人休憩。

The edge of the sidewalk can be set with high and wide bollards for pedestrians seating and resting.

实践案例
Practice

现状 current status

改造方案效果图 Design sketch of reconstruction plan

现状 current status

改造方案效果图 Design sketch of reconstruction plan

123

Chapter Three

珠江新城绿道改善建议

Zhujiang New Town Greenway Improvement Suggestion

概况
Overview

珠江新城位于广州东部新中轴线上，北起黄埔大道，南至珠江河边，西以广州大道为界，东抵华南快速干线。集国际金融贸易、文娱、行政和居住等城市一级功能设施于一体，是广州最现代化的区域，为21世纪广州市中央金融商务区，也是集中体现广州国际都市形象的窗口。

Zhujiang New Town is located in the new central axis in east of Guangzhou.It starts form Huangpu Avenue to the north,Pearl River to the south, Guangzhou Avenue to the west and Huanan Expressway to the east.It combines first class urban functional facilities together such as international finance, trade, commerce, culture and entertainment,administration,as well as residence, which makes the area most modern of Guangzhou.It is the central financial business district of Guangzhou in 21st century, and the window that intensively reflects the image of Guangzhou as an international city.

珠江新城自建设之时起，一直秉承着高标准的建设理念，在区内建设大量的公共设施和公园绿地，如花城广场，已经成为城市的新地标。2010年绿道建设时，珠江新城也第一时间进行了绿道建设。尽管大量高新设施都被运用于该区域，但前来绿道休闲的市民并不多，这是因为在珠江新城内，通过步行或骑自行车的方式到达绿道并不容易。高标准的、便于机动车出行的道路系统，行人和自行车使用起来并不方便。如何让更多的市民更好的使用绿道，是我们现在面临的主要问题。

Since the construction of Zhujiang New Town, it has been adhering to the concept of building a high standard construction. There are public facilities, parks and greenspaces inside the area, such as Flower City Square which has become the new landmark of the city. It constructed the first greenway in 2010. Although a large number of the new facilities are applied in this area, the residences are not many, as it is not convenient for people to reach by walking or cycling. The high standard road traffic system is convenient for motor vehicles, but not convenient for pedestrians and bicycle riders. How to make more people use the greenway is the main problem that we are facing.

实践案例
Practice

125

Chapter Three

2 现状问题分析
Current Condition Analysis

2.1 现状概述
Current Condition Overview

从宏观上来看,造成市民使用绿道不便的原因主要有两个:一是珠江新城内部交通系统设计是基于对机动车使用者需求的设计,方便机动车的快速通行,采用的大街区和长距离过街的设计,不符合行人出行的尺度需求,造成了居民骑行和步行的不便。局部细节设计被忽略和建设、管理的不到位,更是让目前的慢行系统支离破碎,无法连续的原因。二是现有绿道系统的连接不便,无法形成一个高效的系统,例如处于区域中心的花城广场,缺乏自行车道,这就意味着,区内绿道间的交换交通必须依靠区内慢行系统来连接,而区内慢行交通系统的品质并不高。

珠江新城有很好的绿道,但是想骑自行车去体验绿道并不容易
There is a nice greenway system in Zhujiang New Town, but it is not easy for bicycle riders to experience it.

There are mainly two reasons making the current greenways difficult to be used by the public, from the aspect of macro view. Firstly, in Zhujiang New Town, large block and long distance between crossing accomodate the demand of vehicles, but these do not match the scale of pedestrians and bicycles, which cause inconvenience of walking and cycling. Poor detailed design, construction and management cause the discontinuity of the current non-motorized traffic system. Secondly, the existing disconnected greenway system has low effeciency. For example, Flower City Square at the central area is lack of bicyde lane, which means the connection between the greenways in Zhujiang New Town depends on the Town's non-motorized traffic system which, however, is in relatively poor quality.

以花城广场为例，花城广场与花城大道绿道相交，但是，想要骑自行车游览这两条绿道，是无法实现的。这是因为，花城广场没有自行车道。即便是花城大道绿道自身东西两段，也无法实现自行车交通的转换。由于周边慢行系统不发达，行人想要进入花城广场，也仅能通过四个天桥，相当不便。

Using Flower City Square as an example, the square intersects with the greenway on Huacheng Avenue. However, cycling on these two greenways is impossible. That's because there is no bicycle lane planned in Flower City Square. Cycling from the east section to the west section is also impossible. Even the east section and the west section of Huacheng Avenue can not achieve the conwersing of bicycle transportation. Because of the low quality of the non-motorized traffic system in the area, pedestrians can only enter the square through four overbridges, which makes the walking inconvenient.

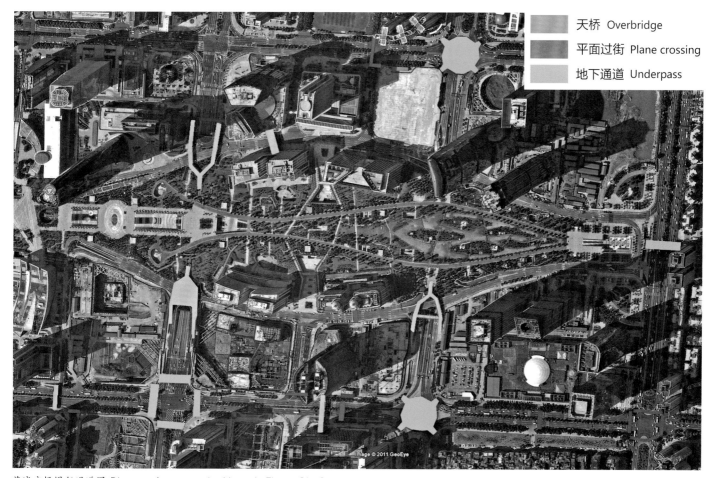

花城广场慢行通道图 *Diagram of non-motorized lanes in Flower City Square*

Chapter Three

2.2 现状典型问题及改善建议
Current Issues and Improvement Suggestions

现状问题 Current Issue

现有绿道间联通不便，很难实现各个绿道间的交通转换。

The disconnection of the existing greenway causes difficulty of traffic conversion between various greenways.

花城大道绿道的东西两段，自行车无法互通，骑行者只能违章使用地下通道。

The bicycle lanes in west and east side of greenway in Huacheng Avenue are not connected with each other, so bicycles can only illegally use underground passages.

花城广场沿线缺乏自行车道。

The Huacheng Avenue is lack of bicycle lane, so bicycles have to ride on the motor vehicles lane.

改善措施 Improvement Suggestion

梳理现有绿道网络，打通关键的节点和路段，在花城广场外侧增设自行车道并连接花城大道自行车道，将绿道整合成一个有机的网络体系。

Reorganize the greenway networks, connect important junctions and sections, and construct bicycle lanes outside Flower City Square which connect with the bicycle lane inside Flower City Square, to combine the greenway into an organic network system.

自行车可使用公共空间。

Bicycle can use public space.

自行车可在广场骑行。

Bicycle can ride in the square.

现状问题 Current Issue

慢行系统网络不全面，无法形成一个整体，也无法提供高质量的服务方便市民到达和离开绿道。

The missing of part of the non-motorized traffic system cannot form a high quality greenway network system for citizens to reach and leave.

自行车车道不连续，骑行者宁可和机动车混行，也不愿使用自行车道。

Disconnection of the bicycle lane; mixed traveling with bicycle riders; and bicycle lanes are not preferred by people.

马路对面即是绿道，但行人和自行车却无法过街，想去绿道必须绕行至下一个路口。

The opposite side is the greenway, but pedestrians and bicycles cannot cross the street. People have to make a detour to the next intersections to reach the greenway.

改善措施 Improvement Suggestion

逐步梳理和完善慢行系统网络，建立适合慢行尺度的慢行系统网络。

Gradually combing and improving the network of slow system, establish the non-motorized traffic line system which is suitable for small scale travel.

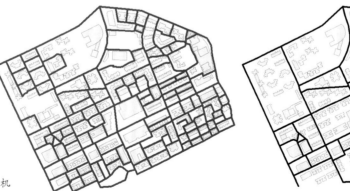

由于慢行尺度较小，高质量的慢行网络要比机动车的网络密集很多。

Since the small-scale of non-motorized travel, a high quality non-motorized network is much denser than a motor network.

N   0 30 60   120   180   240 Meters
— 非机动车与步行网络 Pedestrian and Bicycle Networks
— 机动车网络 Vehicle Network

Chapter Three

现状问题 Current Issue

现有道路断面，只注重慢行系统与机动车道的分离，并未设置物理分隔的自行车道和人行道。这种道路断面设计，在实际使用中并不能很好的起到分离自行车流和行人流的作用。而一些道路断面的不合理设计，导致了慢行系统、特别是自行车道的中断。

Bicycles and pedestrians are mixed in road section with no physical separation. In practical use, some unreasonable designs cause interruption of non-motorized traffic system, especially in bicycle lane.

道路断面功能设计不合理，自行车道空间被休闲椅占用。
Bicycle passage space is occupied by public benches.

行人行走在自行车道上，自行车走在人行道上。
Pedestrians on bicycle lane, bicycles on sidewalk

改善措施 Improvement Suggestion

对规划区内未设置自行车道的道路，建议未来如有改造机会，尽量采用分离式的设计，通过功能区的划分，分离机动车道、非机动车道和人行道。调整现有道路断面，通过设置分隔带等多种方式，有效分隔人行道、自行车道。

For the roads without bike lane, it is suggested to have clearly separated function area in the future when there is opportunity for rennovation. The cross section can be changed with additional physical separation to define walkway and bike lane.

道路空间充足时，可设置分隔带。
When the road space is enough, the separation zone can be set to completely separate various functions.

道路空间受限时，可使用挡块分隔。
When the road space is limited, blocks can be used as separation

实践案例
Practice

现状问题 Current Issue

慢行系统空间被侵占，无法形成一个连续的系统。

The non-motorized traffic space is occupied, thus cannot remain as a continuous system.

自行车道被公交站占用。
Bicycle lane is occupied by bus station

现有自行车道完全被报刊亭和自行车停车占用。
The existing bicycle lane is completely blocked by the book stalls and parking cars.

未设置隔离，自行车道和人行道被停车占用。
Separation zone is not planned, so the bicycle lane and side walk are occupied by illegally parking.

改善措施 Improvement Suggestion

对慢行系统被占用的情况，像报刊亭等设施，建议逐渐清理，还原道路应有的功能；对于机动车停车占用慢行通道的情况，则建议逐渐完善，在路侧设置护柱，将机动车彻底隔离在人行道外围，而对于公交站，则可以调整公交站的位置，保证慢行系统有足够的使用空间。

As for the occupation of book stall on non-motorized traffic system, it is suggested that gradually clean it and recover the original function of the road. As for motor vehicles' parking, it is recommended to gradually improve the roadside bollards and completely isolate the motor vehicles. Guarantee enough space of non-motorized traffic system by the readjusting of bus stations.

在路侧设置护柱，阻止机动车进入慢行系统。
Set bollards at the edge of sidewalks to prevent vehicle entering the non-motorized traffic area.

公交港湾停靠站的设置必须以保证足够的慢行空间为前提。
The setting of bus station should ensure the continuity of non-motorized traffic space.

Chapter Three

现状问题 Current Issue

珠江新城行人和自行车目前过街不便，不仅是现有通道难以保障行人和自行车过街、自行车道在交叉口中断，同时也缺乏有效过街通道，行人过街缺乏保护等情况较普遍。

Pedestrians and bicycles are inconvenient to cross the streets at Zhujiang New Town, not only through the existing passages with bicycle lanes interrupted at intersections, but also lack of effective crossing passage, which causes unsafety for pedestrians' crossing.

改善措施 Improvement Suggestion

完善现有过街设施，增设自行车专用过街通道和路中安全岛，并在过街需求集中处增设路中过街通道。

Enhance the existing pedestrian crossing facilities, at the same time, add bicycle crossing passage and road safety island at intersections and construct street crossing passage at the places where have concentrated demand for street crossing.

交叉口缺乏人行过街通道。
No sidewalk passage at intersection.

专用的自行车过街通。
Bicycle street crossing passage.

缺乏过街通道。
Residents have street crossing demand but lack of crossing passage.

未设置行人安全岛。
No setting of pedestrian safety island.

良好的过街通道。
Qualified crossing passage.

实践案例
Practice

现状问题 Current Issue

机动车出入口的处理不好，自行车道、人行道被机动车出入口切断。骑行者骑行不便，以至于骑行者宁愿在机动车道上骑行，也不使用自行车道。

Irrational design of the entrance and exit of motor vehicles interrupts bicycle lane and sidewalk. Bicycle riders rather ride on motor vehicles line than ride on bicycle lane.

改善措施 Improvement Suggestion

保障自行车和行人的路权，改善现有机动车出入口的设置。建议根据情况，逐步实施改造升级，抬升机动车出入口处的自行车道及人行道，保持慢行系统的通畅。

Guarantee the right of bicycles and pedestrians, and improve the existing vehicle entrance layout. We suggest lift the entrance of the motor vehicle and the sidewalk, to maintain the smooth flow of the system.

慢行系统被机动车出入口切断。

The non-motorized traffic system is interrupted by the entrance and exit of motor vehicles.

抬升机动车出入口，保障自行车和行人的路权。

Uplift the entrance of the motor vehicles and guarantee the right of non-motorized traffic.

在机动车出入口处，自行车根本无法骑行。

Bicycle cannot pass as there are many entrances and exits of motor vehicles.

在小路口，也可以采用同样的设置。

Same settings can be adopted at the small intersections.

Chapter Three

总体规划布局
Overall Layout

3.1 完善绿道网络
Improve Greenway Network

作为珠江新城绿道系统的中心区域——花城广场，缺乏自行车道不仅会造成绿道与绿道间骑行的连接不便，一定程度上还遏制了市民在区内绿道上骑行的意愿。

The lack of bicycle lane in Flower City Square, as the central area of the greenway system in Zhujiang New Town, will not only make inconvenient connection among the greenways, but also hinders citizens riding on greenways, which causes inconvenience of regional bicycle riding.

现有花城广场两侧的珠江大道机动车空间较宽，且机动车流量不大，可移除花城广场最内侧的机动车道，并设置专用的自行车道。

The existing motor traffic space on Zhujiang Avenue at both edges of Flower City Square is large, but the traffic flow is relatively small. The lanes next to Flower City Square can be turned into dedicated bikelane.

实践案例
Practice

黄埔大道仅有一个右车道，但在珠江西路上设置了3车道提供给黄埔大道的右转车辆，存在着浪费。

There is only one right lane in Huangpu Avenue, but on Zhujiangxi Road, there are 3 lanes provided for the motor right turning vehicles. It is a waste of space.

珠江西路上的机动车不多，自行车会选择在机动车道上骑行。

The motor vehicles on Zhujiangxi Road are not many, so bicycles choose to ride on motor vehicle lane.

花城广场可建成高品质的共享空间

Flower City Square can be constructed as a high quality shared space

Chapter Three

图中所示绿线为建议增设自行车道位置，在北部建议借用内侧机动车道，设置专用自行车道，南侧建议在花城广场内部设置自行车道。

The green lines shown in the image is the position of suggested bicycle lanes. It is suggested that use the motor lane of inner side in the north as bicycle lane. The southern bicycle lane is suggested to be set inside Flower City Square.

花城广场北部建议增设自行车道位置图。

The position of suggested bicyde lanes in the north of Flowe City Square.

规划建议移除珠江西路、珠江东路最内侧的一条机动车道，并设置成专用的自行车道，完善整体绿道系统网络。

Plan on using the inner side of the motor vehicles lane on Zhujiangxi Road, and Zhujiangdong Road, to set dedicated bicycle lanes and improve the overall greenway system.

花城广场南侧自行车道建议延伸至花城广场内部，利用花城广场高品质的公共空间，采用与行人混行的模式，保持自行车道的连续。

It is suggested to extend the bikelane south of Flower City Square into the Square, sharing with pedestrians to keep bikelane continuous.

实践案例
Practice

花城广场中部建议增设自行车道位置图。

The position of suggested bicyde lanes in the central City Square.

花城广场南部建议增设自行车道位置。

The position of suggested bicyde lanes in the south of Flowe City Square.

Chapter Three

3.2 行人轨迹调查
Pedestrian Trajectory Survey

为更好地了解实际的行人过街流量及过街位置分布情况，获取周边居民进入花城广场的路径，我们对珠江西路道路进行了路段行人轨迹调查。行人轨迹调查就是通过记录道路上每个过街行人的步行过街轨迹，获得道路行人过街流量及过街位置分布等的资料，为现状道路的改造提供依据。

For a better understanding of the actual pedestrian flow and distribution, we conducted a survey about the passages where the surrounding residents choose to enter the Flower City Square and also give a survey on pedestrian trajectory on the road section of Zhujiangxi Road. Pedestrian trajectory survey means to obtain the first-hand sata on pedestrian flow through recording every pedestrian's trajectory on the road, to provide basis for the road transformation.

注：图中数据来源于 ITDP 在 2011 年 6 月的现场调查。

Note: the data in the figure comes from the field survey of ITDP in June 2011.

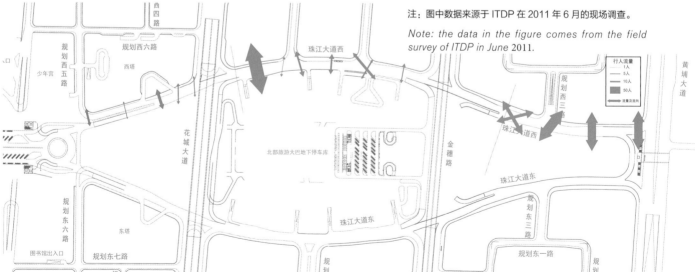

珠江新城珠江西路行人过街轨迹。 *The pedestrian trajectory in Zhujiangxi Road in Zhujiang New Town.*

实践案例
Practice

注：图中数据来源于ITDP在2011年6月的现场调查。

Note: the data in the figure is from the field surrey of ITDP in June 2011.

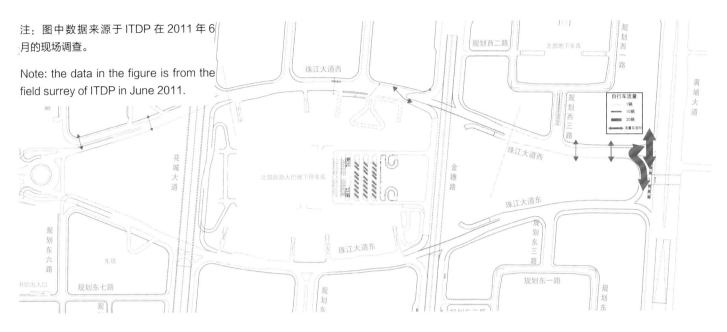

珠江新城珠江路自行车过街轨迹。The bicycle trajectory on Zhujaingxi Road in Zhujiang New Town.

尽管珠江西路尚未设置自行车道，但是仍有自行车骑行的需求，而且存在着几处集中的过街区域。

Although bicycle lane is not constructed in Zhujiangxi Road, there are still demand of riding bicycles. Moreover, there are a few concentrated pedestrian crossing areas exist.

Chapter Three

结合道路行人过街轨迹的分析，完善珠江西路过街设施，重新界定珠江西路行人及自行车过街通道，并对其进行相应的改造设计，提供更符合市民出行习惯的过街通道，为市民进出花城广场提供便利。

Combine the pedestrian crossing trajectory to improve the crossing facilities in Zhujiangxi Road. Through the pedestrian crossing survey, obtain pedestrian and bicycle's street crossing passage to make relevant transformation design, which provide citizens a crossing passage based on their traveling habits and also provide convenience for the citizens entering into the Flower City Square.

改善措施：

Improvement measures:

· 在珠江西路最内侧增设自行车道，并设置分隔带，分离自行车道和人行道；
· 在行人集中过街处新增过街通道；
· 在自行车过街处设置独立的过街通道；
· 增设交通导流岛；
· 人行道在道路交叉口处向外延伸，缩短行人过街距离。
· 增设信号灯。

· Add a bicycle lane to the inner side of Zhujiangxi Road and equip with separation belt to isolate bicycle lane and pedestrian lane.
· Add street crossing passage at the places where pedestrians crossing is concentrated.
· Add independent bicycle crossing passage.
· Add transportation diversion islands.
· Extend the intersections and shorten the distance of pedestrian crossing.
· Add signal lamp.

花城大道绿道现状东西两段无法互通，行人可通过使用天桥连接，但自行车就无法互通，目前花城大道的自行车使用者只能违章使用机动车隧道通过花城广场。

The east and west section of Huacheng Avenue greenway cannot link to each other. Pedestrians' passage can be connected through bridges, but bicycles lane cannot link to each other. Currently, the bicycle users in Huacheng Avenue can only illegally pass through the motor vehicles' tunnel in Flower City Square.

花城大道绿道的东西两段自行车是无法互通的，自行车只能违章使用地下通道。

The east and west side of the greenway in Huacheng Avenue cannot link to each other, and bicycles can only illegally pass through the tunnel underground.

为构建珠江新城慢行系统的骨干网络，保证其绿道的连续性，需连接花城大道和金穗路的绿道，可在花城广场内设置自行车通道，采用行人和自行车混行的模式，具体路径可按照可实施性选定。

For the construction of the non-motorized system in Zhujiang New Town and the continuity of the greenway, we need to connect Huacheng Avenue and Jinsui Road. Bicycle passage can be planned in Flower City Square, and we can adopt pedestrian and bicycle mixed mode. The specific passage can be made in accordance with the implementation.

Chapter Three

3.3 构建密集的街道网络
Build Dense Street Network

对于城市交通而言,并非道路越宽就越通畅、效率越高。实际上,由宽阔道路划分出的大街区使中国的道路拥堵状况愈发严重。案例研究表明,较密集的狭窄道路网络更有利于交通通畅,同时能创建更多直达路线并提高行人安全性。道路设计应实现乘客机动性的最大化,而不是车辆通行能力的最大化。在宽度较窄的道路上实行单向机动车行驶,并可骑车、步行,可以缩短交通信号灯延迟,从而极大地缓解道路拥堵状况,并降低机动车燃料消耗。

For urban traffic, wide roads do not necessarily mean smooth and effective traffic. In fact, dividing the blocks with wide roads has caused worsening traffic jam. A case study shows that a dense network with narrow roads is better for traffic flow, and it can create more direct walking routes and improve the safety of pedestrians. The road design should maximize the mobility of passengers rather than maximize the capacity of vehicles. On the road of narrow width, one-way vehicle lanes can be applied, and it is better for riding and walking. It can also shorten the delay by traffic lamp, and greatly mitigate the traffic jam and reduce energy consumption.

创建密集的慢行网络,符合慢行交通的出行尺度,可有效改善步行和自行车的出行环境,鼓励市民步行或使用自行车。

Creating a concentrated non-motorized network, and make it in line with the traffic travel scale can effectively improve the environment of walking and bicycle traveling and encourage people to walk or use bicycle.

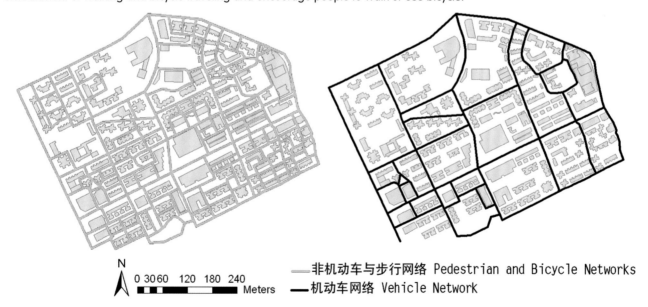

江南西小区步行及自行车道网络图 *The sidewalk and bicycle lane network diagram in Jiangnanxi Neighborhood*

实践案例
Practice

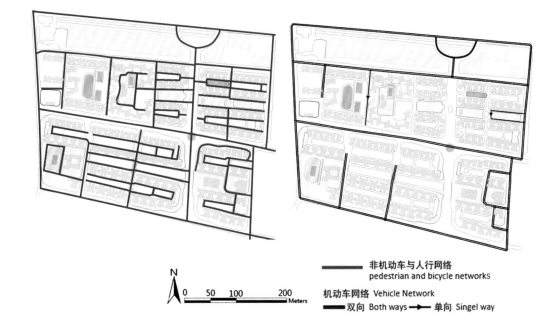

骏景花园步行及自行车道路图路。
The sidewalk and bicycle lane network diagram in Junjing Garden.

非机动车与人行网络 pedestrian and bicycle networks
机动车网络 Vehicle Network
双向 Both ways　单向 Singel way

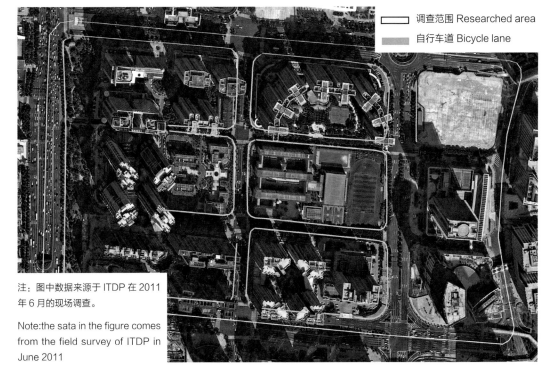

调查范围 Researched area
自行车道 Bicycle lane

注：图中数据来源于 ITDP 在 2011 年 6 月的现场调查。

Note: the sata in the figure comes from the field survey of ITDP in June 2011

珠江新城小区内慢行系统并不密集，很多地方不完善，如图所示，区内自行车道不连续，从区内到花城广场，缺乏直接连接的自行车道。

The non-motorized traffic system in Zhujiang New Town is not concentrated. Many places are not completed. As shown in the figure, the bicycle lane inside the area is not consistent, and lack of directly connected bicycle lane from the area to Flower City Square.

143

Chapter Three

4 道路横断面改善建议
Road Cross Section Improvement Suggestion

4.1 道路断面现状
Current Condition of Road Section

为保持骑行和步行的良好的体验感，应保持慢行系统的连续，并且设置隔离的慢行系统，分离机动车道、自行车道和人行道。

In order to maintain a good experience of riding and walking, we should keep the continuity of non-motorized traffic system, and set isolated non-motorized traffic to separate motor vehicles, bicycles and pedestrians.

珠江新城的道路断面，绝大多数的断面中都做到了慢行系统与机动车道的分隔，而对于人行道和自行车道，并未实现物理分离，只是通过颜色铺装的来区分。

Most of the road section in Zhujing New Town have made separation of non-motorized traffic system and motor vehicles. Sidewalks and bicycle lanes, however, are not physically separated. They are only divided by colors.

珠江新城内部道路断面图（部分）
Part of the road cross section inside Zhuajiang New Town

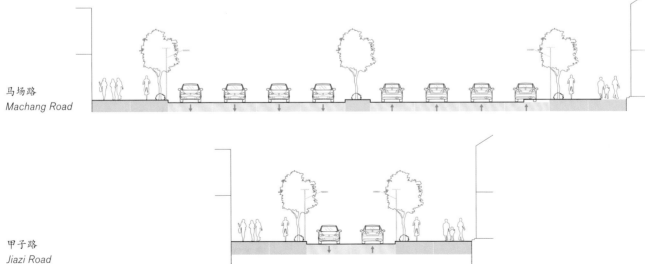

马场路
Machang Road

甲子路
Jiazi Road

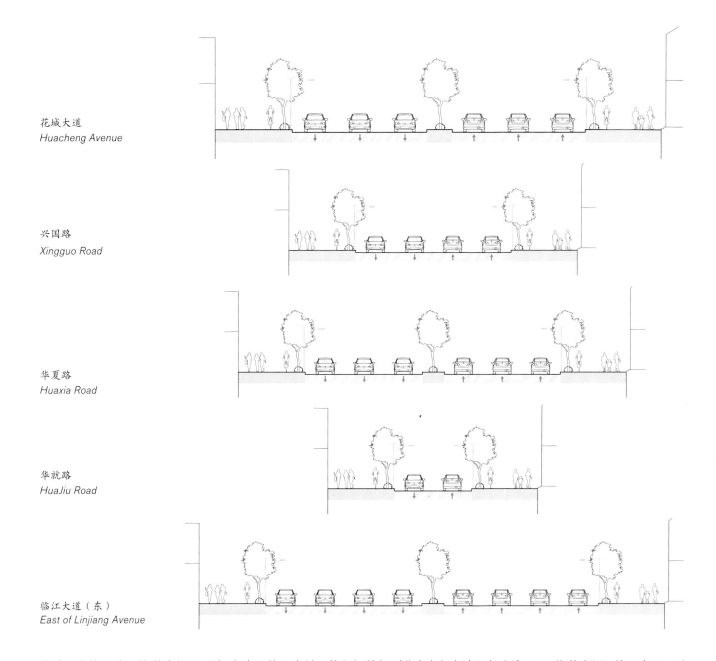

花城大道
Huacheng Avenue

兴国路
Xingguo Road

华夏路
Huaxia Road

华就路
HuaJiu Road

临江大道（东）
East of Linjiang Avenue

这种没有物理分隔的道路断面设计，在实际使用中并不能很好的起到分离自行车流和行人流。而一些道路断面的不合理设计，导致了慢行系统，特别是自行车道的中断。

The design with no physical separation cannot divide bicycle flow and pedestrian flow in practical use. Some unreasonable design causes interruption of non-motorized traffic system, especially bicycle lane.

Chapter Three

4.2 道路断面改善建议
Road Section Improvement Suggestion

为了保持珠江新城慢行系统的连续，满足市民高质量的出行需求，针对现状存在的，阻碍慢行系统连续的典型性问题，制定了改善性建议。

To keep the continuity of the non-motorized traffic system in Zhujiang New Town, and to satisfy residents with high quality traveling experience, we make improvement suggestion on current typical issues about the continuity of non-motorized traffic system.

本建议仅为概念性建议，在具体实施时，应根据各道路的功能、道路实际宽度、现状使用情况等相关因素，结合本次改善建议，共同制定出有针对性的改善方案。

This suggestion is only a conceptual proposal. In practical implementation, it should be based on the function of the road, the actual width of the road, the status and other related factors. Combining with the improvement suggestion, we specifically make improvement program.

临建筑区	行人通行区	路中小品	自行车道	分隔带
Area adjacent to building	*Pedestrians passing area*	*Decoration*	*Bicycle lane*	*Separation belt*
分隔建筑与人行道，提供缓冲空间。	路边主要的人行通区，无障碍通道设计可为行人步行提供高质量的步行空间。	分隔相邻的人行通道和自行车道，为其提供缓冲。	自行车专用道，为自行车设计的独立空间。	路边分隔带，分隔自行车道和机动车道。
Separate buildings and sidewalk to provide cushion space.	*Roadside pedestrians passing area. The design of barrier free provides pedestrians with high quality walking space.*	*Separate adiacent sidewakl and bicycle lane to provide cushion space.*	*The bicycle lane is the independent space designed for bicycles.*	*Roadside separation belt separates bicycle lane and motor way.*

路侧空间临建筑区、行人通行区、路中小品、自行车道分隔带划分，应根据道路的实际情况，作出有针对性的设计，并符合现状及未来规划的区域和街道沿线的土地利用情况。

The separation of roadside temporary building area, pedestrians passing area, decoration and bicycle lane should according to actual situation to mak specific design, and accord with current and future planning in the area and along the street.

非机动车交通系统设置要点：
· 保障安全，空间上分离人行道、自行车道和机动车道；
· 非机动车交通优先于机动车的路权设计；
· 足够的慢行空间；
· 宁静交通设置；
· 合理设置道路家具；
· 无障碍设施的设置。

Key point of non-motorized vehicle traffic system:
· Security, space for separation of sidewalks, bicycle lane and motor vehicle lane.
· Non-motorized traffic system should have the priority over motor vehicles.
· Enough space.
· Quite traffic settings.
· Reasonable setting of road furnishing.
· Barrier free facilities.

由于珠江新城的人行道和自行车道普遍被设置在一起，并且目前很难对现有道路进行大规模的改造，因此，根据道路的不同情况，提出以下几种改善建议：
路侧空间不足，机动车道有空间，且流量不大，建议移除一条机动车道，并设置成专用自行车道。

Because of the setting of same plane of sidewalk and bicycle lane is quite common in Zhujiang New Town, according to the different road condition, the following suggestions can be put forward for improvement.
Not enough roadside space, there is large room for motor vehicles but the flow is small. It is suggested to remove one motor vehicle lane and set up a bicycle lane.

华夏路 - 现状
Huaxia Road-current

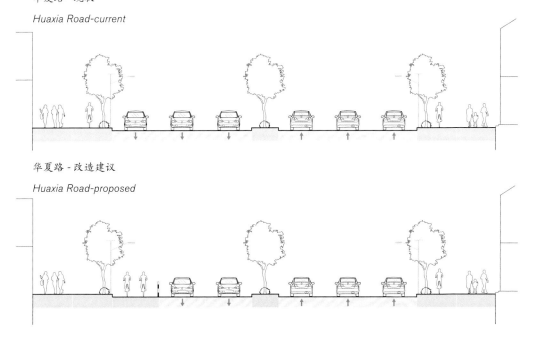

华夏路 - 改造建议
Huaxia Road-proposed

Chapter Three

马场路 - 现状

Machang Road-current

马场路 - 改造建议

Machang Road-proposed

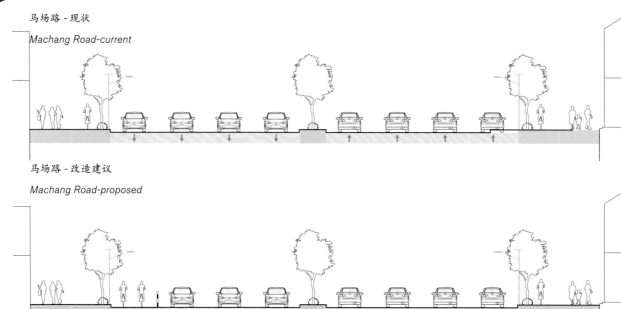

路侧空间充足的情况下，建议增设分隔带、护栏或分隔挡块分离人行道和自行车道。

When roadside space is quite large, it is suggested to add separation belt, bollard and isolation block to separate sidewalk and bicycle lane.

花城大道 - 现状

Huacheng Avenue-current

花城大道 - 改造建议

Huacheng Avenue-proposed

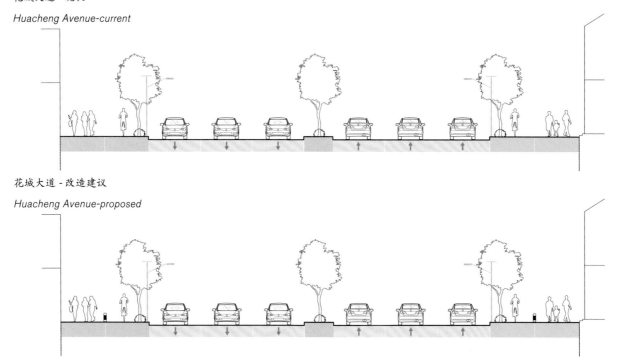

实践案例
Practice

兴国路 - 现状

Xingguo Road-current

兴国路 - 改造建议

Xingguo Road-proposed

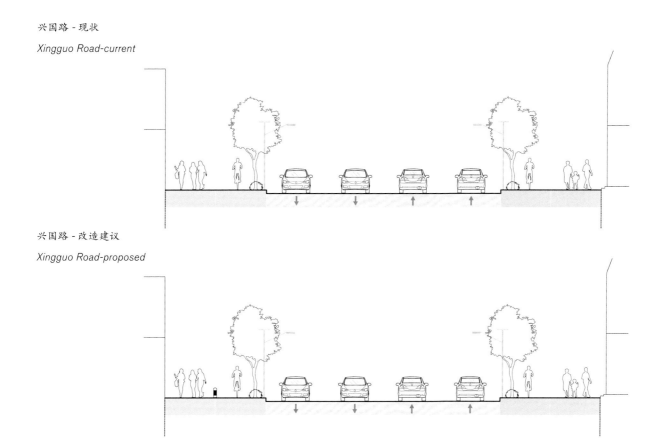

小支路，无自行车道，建议仅从交通管理角度入手，允许自行车与机动车混行，共享道路空间，无需进行改善型设计。

For small branch road without bike lane, traffic management should be enhanced to allow bikes and vehicles share the street, reconstruction design is not necessary.

华政路

Huazheng Road

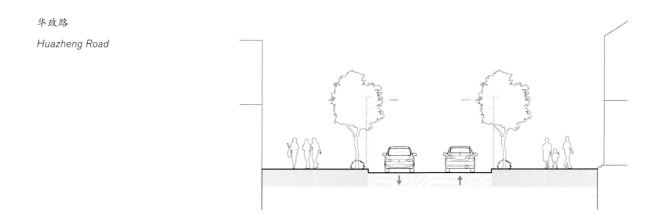

149

Chapter Three

5 过街通道改善建议
Street Crossing Improvement Suggestion

5.1 过街通道现状分析
Analysis of the Current Street Crossing Situation

过街通道在交通系统中十分重要，它的对象包含所有的使用人群——当地居民和游客，需要能便利地到达并能链接到周边地区。此时交叉口和过街通道的设置就变得十分重要，不仅要让过街变得方便安全，同样也要符合多数人的过街习惯和保证区域交通运行的顺畅。

Street Crossing is very important in transportation. Its object contains all the people, from residents to tourists, who all need to reach the surrounding areas easily. It can not only make street crossing convenient and safe, but should also in accordance with most people's crossing habits and ensure the smooth operation of regional traffic.

珠江新城的过街系统并不是很完善，在一些主要的路口甚至缺乏人行过街设施。在一些主要的交叉口，设置了花费巨大的地下通道，但是却没有达到预期的效果。

The crossing system in Zhujiang New Town is not perfect. In some main junctions there is even no pedestrian crossing facility. In some main intersections, the setting is to construct underground passage, which spends a large amount of money, however, still cannot meet the prospective results.

下页为珠江新城现状通道分布总图，将珠江新城分为1、2、3不同区域，可更加清晰地看到区域的过街通道情况。

The page below is the channel distribution map in Zhujiang New Town, and it is divided into different regions of 1, 2, 3. We can see the whole area's crossing condition.

下页图中蓝色区域为金穗路、广州大道、花城大道、花城广场围合的地块，尽管在总图中此区域的过街通道还是比较密集，但实际调查显示该区域过街通道仍有较多的缺失。

The blue area constrains Jinsui Road, Guangzhou Avenue, Huacheng Avenue and Flower City Square. Although in general, drawing the crossing passage in this area is concentrated, when zooming in the local graph, we can see that there are still missing a large number of street crossing.

实践案例
Practice

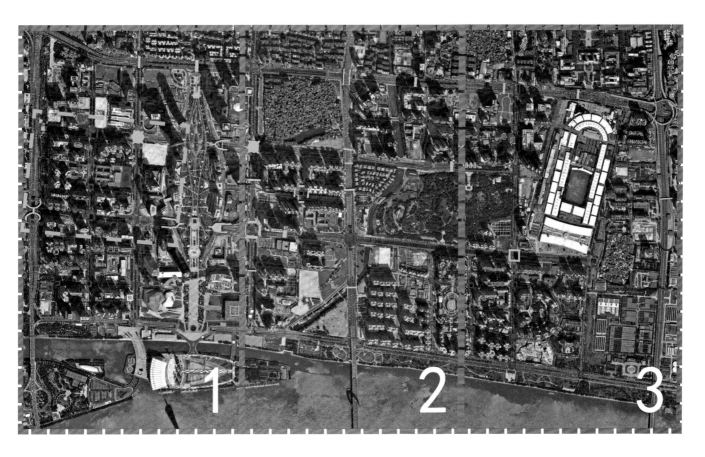

珠江新城过街设施分布总图
Street crossing facilities in Zhujiang New Town

天桥 Overbridge
平面过街 Plan crossing
地下通道 Underpass

Chapter Three

路中过街通道 Mid-Block Crossing

为保证珠江新城的机动车高效的运行，在多数路段均未设置过街通道，给市民的出行带来了不便。周边居民为了能方便地过街，甚至在中央分隔带中踩出了一条"路"。

In order to maintain the high efficiency of the motor vehicles moving in Zhujiang New Town, most of the road sections are not equipped with crossing passage. It causes inconvenience for the residents. In order to facilitate the needs of crossing, the surrounding residents trample out a "road" in the central zone.

过街小道
Crossing path

中央分隔带上的护栏被拆除。
For the convenience of street crossing, residents removed the bollard in the central zone.

花城广场东西两侧无法通过，自行车想通过必须从隧道穿越，与机动车混行。

The east and the west sides of Flower City Square are impassable, so bicycle riders have to mix with motor vehicles and pass through the tunnel.

5.2 平面过街和立体过街
Crosswalk and Pedestrian Bridge or Tunne

珠江新城中建设的高标准的过街通道，实际使用率并不高。市民过街时更多地愿意选择平面过街。

High-standard crossing passages in Zhujiang New Town are, in fact, not frequently used. People prefer use surface crosswalk.

金穗路 – 华夏路交叉口过街 Jinsui Road-Huaxia Intersection Crossing

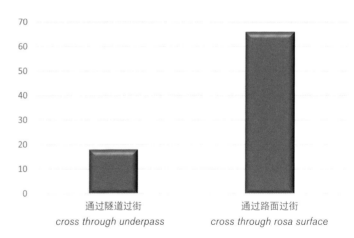

通过隧道过街　　　　　通过路面过街
cross through underpass　　　*cross through rosa surface*

15分钟内道路交叉口的过街流量
Crossing rate at intersection within fifteen minutes.

Chapter Three

在对金穗路－华夏路交叉口西北角路口行人过街的调查中，15 分钟几有 66 人通过地面过街，18 人通过隧道过街；其中，自行车全部通过地面过街。

The survey of the pedestrian crossing in north west corner of intersection in Jinsui Road to Huaxia Road shows that 66 people cross the street through the ground within fifteen minutes, and 18 people cross the street through tunnel, and for bicycles, they all pass through the ground.

花城广场入口过街 Crossing at the entrance of Flower City Square

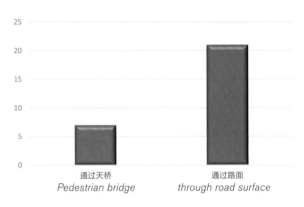

15 分钟内进入花城广场的过街流量

Crossing rate at entering Flower City Square within fifteen minutes.

在调查的 15 分钟内，进入花城广场的 29 个行人中，有 8 人选择使用天桥进入，其他 21 人选择违规横穿马路，通过地面过街的形式进入花城广场。

And in the population flow of entering Flower City Square within fifteen minutes, there are 8 people choose to use the bridge, 21 choose illegally cross the roadway.

从现状过街天桥和隧道的使用情况来看，多数市民并不喜欢使用天桥或隧道过街。在条件允许的路口，市民宁可选择违章从地面过街，也不愿使用天桥或隧道。在过街通道的设计时，对市民的需求应予以足够的重视。

From the current situation of the footbridge and underground tunnel, most of the citizens do not prefer footbridge or tunnel. Citizens would rather choose illegally cross the road than using footbridge and tunnel. In the design of street crossing, it should be paid enough attention.

实践案例
Practice

过街通道在路口的不当设置，导致了慢行系统的中断。
The improper setting of passage at the intersection will cause some problems, leading to the interruption of non-motorized sytem.

自行车道被中断。
Bicycle lane is interrupted.

无专用自行车过街通道，交警要求骑行者过街时下车推行。
No bicycle lane crossing passage, the traffic police need to ask the riders to get off the bicycle when crossing the street.

Chapter Three

5.3 交叉口改善建议
Intersection Improvement Suggestions

过街通道设计要点 Key Points of Crossing Design

为了更好地满足区域内慢行交通使用者的需求，交叉口的设计也应突出为非机动车优先交通服务的特点，在交叉口设计时应注意以下要点：
· 降低机动车在交叉口和人行过街处的车速；
· 符合多数人的过街需求；
· 减少行人暴露在机动车流中的危险；
· 提高机动车驾驶员的可预见性；
· 行人及自行车过街优先。

The crossing design should better meet the demand of non-motorized traffic users within the area. The intersection design should highlight the priority of non-motorized transportation. The following key points shall be taken into account.
· Reduce the speed of the motor vehicles at intersection and pedestrian crossing .
· Accomodate the crossing demand of the majority of pedestrians.
· Reduce the risk of pedestrian exposing in the traffic flow.
· Improve the predictability of motor vehicle drivers .
· Prioritize pedestrian and bicycle crossing.

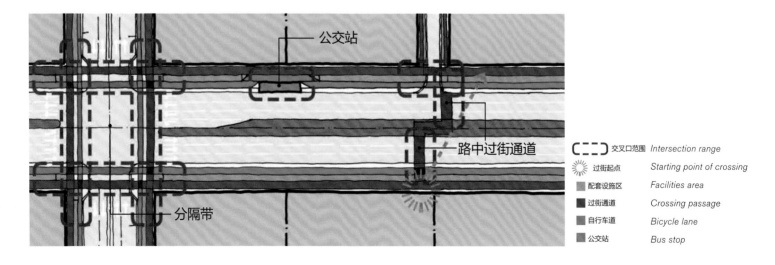

实践案例
Practice

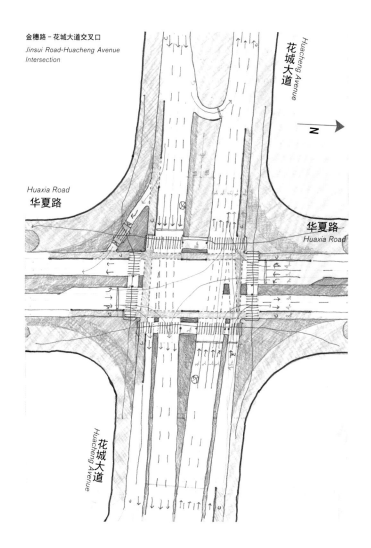

金穗路-花城大道交叉口
Jinsui Road-Huacheng Avenue Intersection

华城大道 Huacheng Avenue

华夏路 Huaxia Road

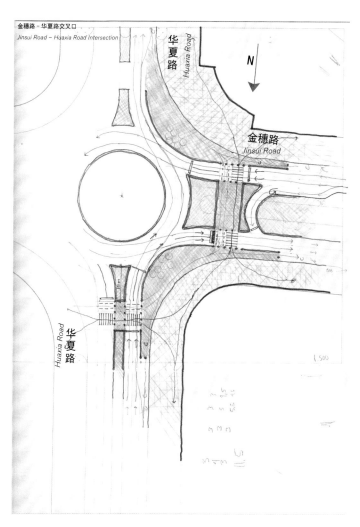

金穗路-华夏路交叉口
Jinsui Road – Huaxia Road Intersection

华夏路 Huaxia Road

金穗路 Jinsui Road

华夏路 Huaxia Road

Chapter Three

6 配套设施建设
Facilities Construction

6.1 自行车停车点
Bicycle Parking Point

在珠江新城，绿道及绿道沿线并未设置自行车停车点，甚至在珠江新城内部道路沿线，也没有自行车停车点。区内停车不方便，减少了居民使用自行车在绿道上骑行的意愿，自行车的停放也较为随意。

In Zhujiang New Town, there is no bicycle parking point in greenway and on the road along the greenway. There is even no bicycle parking point inside the inner road in Zhujiang New Town. Parking inside the area is not convenient, and hinders residents riding on greenway. The bicycles are also randomly parked.

在临街区域，停放了48辆自行车。

There are 48 bicycles parking in the area near street.

在公交车站有自行车停车的需求。

There is a need of bicycle parking at the bus station.

自行车的违章停车，已严重干扰了自行车道的正常使用。

Illegal bicycle parking has seriously interrupted the bicycle lane.

在商铺前，也有自行车的停车需求。

There is bicycle parking demand in front of the store.

实践案例
Practice

为了解珠江新城自行车停车难得现状，我们对珠江新城的自行车停车供给和需求进行了调查，从调查的结果来看，目前珠江新城自行车停车缺乏公共自行车停车场，自行车停放较为随意。由于花城广场没有自行车道，在其两侧也没有自行车停车需求，反映出目前基本没有人使用自行车抵达花城广场游玩的状况。

To understand the status of bicycle parking in Zhujiang New Town, we have conducted a survey for the bicycle supply and the needs of bicycle parking. From the survey, the current Zhujiang New Town has no public bicycle park, and the bicycle parking is quite random. There is no bicycle lane in Flower City Square, and there is no bicycle parking needs at both sides of the road. It reflects that there is no people visiting Flower City by bicycle.

规划选定区域自行车停车需求分布图

Distribution graph of bicycle parking demand in regional Zhujiang New Town

Chapter Three

能够骑自行车到达绿道将会是一件令人愉快的事情。因而在绿道附近配套易于使用的自行车停车点变得尤其重要。停车点应有高度的安全保障。同时，自行车停放站应设置在显眼位置，晚上设置良好的灯光照明，同时设置雨篷以防天气的不测。同样重要的是，停车系统的外观美感总体上应给人们一种良好的体验。

It would be a pleasant thing to get people to ride a bicycle to the greenway. And the convenience of facilities is particularly important for the parking point which near the greenway. At the same time, bicycle parking station should be transparent or can be easily observed. It should have a good lighting system in the night, and should have awning to prevent the unpredictable weather. The outer appearance is also important, and it should give people a sense of beauty.

便利的自行车停放点应注意：
· 便利到达和舒适度
· 位于绿道、公园附近
· 具备良好的品质和容纳能力
· 保障自行车和骑行者的安全
· 自行车的停放架和雨篷的高品质设计

The key point of convenient bicycle parking point :
· convenient to reach and comfortable experience .
· near greenway and parks.
· good quality and large capacity.
· safety and protection of bicycles .
· bicycle parking frame and the design of high quality awning .

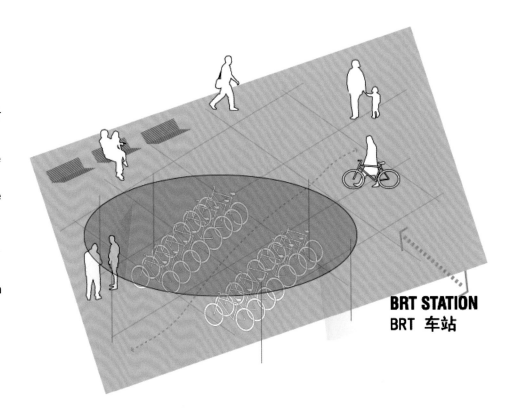

BRT STATION
BRT 车站

根据目前的需求、未来的出入口位置和对外通道的格局，规划在公园对外通道处规划 6 处自行车停车点，自行车停车点设置的规模与未来的交通需求相匹配，未来站点建设根据实际需求量的大小相匹配，分以下三种形式：

According to the current demand on the future entrance and the outward layout, we plan 6 bicycle parking stations in the outer passage of park. The size of the bicycle parking stations should in accordance with the future transportation demands. The construction of the future station should be based on actual demands. They can be classified into 3 forms:

高停车需求。
High parking demand

中等停车需求。
Medium parking demand

低停车需求。
Low parking demand

Chapter Three

根据地块停车情况,规划新增6处自行车停车场,其中3处自行车停车场布置在现状停车集中区域,其余,均设置在花城广场入口处,服务未来使用绿道进出花城广场的市民。

According to the current parking condition, there are 3 bicycle parking stations among the newly constructed 6 bicycle parking stations in the center area, and other parking stations are mainly constructed in the entrance of Flower City Square. They will provide service for the citizens in and out of Flower City Square.

在停车需求较大的区域,并不是必须采用停车楼的形式,也可以采用路边停车带的布置方式。

In the area with large parking needs, the parking building is not necessarily applied; instead, roadside parking can also be adopted.

路边停车场同样可以有大的停车容量

Roadside parking can also have the large parking capacity

路边停车点 Roadside parking point

而对于花城广场出入口处的停车场,可根据未来的实际需求量的大小,利用天桥下空间或采用立体式自行车停车场

As for the parking lot at the entrance of Flower City Square, one can make it according to the actual amount of needs in the future or use the space under the bridge for dimension bicycle parking lot.

利用天桥下空间停车 Use the space under the bridge as a parking space

采用地下停车场 Use underground parking

6.2 公共自行车
Public Bicycle

如何吸引市民到珠江新城休闲并使用绿道？除了提高其可达性之外，还可以优化珠江新城内部及周边路网、改善道路断面及关键节点的设计、加强配套设施的建设能切实方便市民出行和提高出行体验，使市民在珠江新城步行或者骑自行车成为一种惬意的享受。

How to attract people to use the greenway in Zhujiang New Town? IN addition to improve the accessibility, we can also optimize the inner and the surrounding road network and improve the design of road section and key junctions. In addition, strengthening the construction of facilities can effectively facilitate citizens' traveling and enhance traveling experience, and make cycling or walking in Zhujiang New Town a comfortable enjoyment.

公共自行车系统作为一种便民利民的出行方式，能有效缓解公交"最后一公里"问题，与公交、地铁形成大的公共交通网络，提供居民出行"点到点"的服务，同时也为到珠江新城休闲娱乐的市民提供更便捷的出行方式。

Public bicycle system as a way to provide convenience to citizens' traveling, can effectively alleviate the "last mile" problem, and form a great public transportation network with bus and subway. It can provide residents with different transportation services, at the same time, provide more convenient traveling ways for the recreation in Zhujiang New Town.

广州某 BRT 站点旁的公共自行车服务点
Public bicycle service point at BRT station in Guangzhou

里昂某广场上的公共自行车服务点
Public bicycle service point in a square in Lyon

Chapter Three

公共自行车服务点主要分为3类：公交点、公建点及城市内兴趣点、居住点。

Public bicycle service points are mainly divided into 3 categories, which are bus points, public buildings and urban interesting site, and residential areas.

公交点：在公共交通站点、BRT及地铁站附近设置公共自行车服务点，实现公共交通的换乘与接驳。

Bus stops–realize constructing public bicycle service point for transfer and connection with public transportation at the bus station, BRT and subway station.

公交点—与BRT的接驳
Bus station--connected with BRT

公建点及城市内兴趣点：在城市内商业区、广场、学校及公园等公建点及城市内的兴趣点附近设置公共自行车服务点，满足上/下班、上/放学及休闲购物的"最后一公里"出行需求。

Public buildings and urban interesting sites–constructing public bicycle service point in city business areas, squares, and near the schools and parks and other public constructions. Therefore, satisfy people's "last mile" demands on going and off work or school, and leisure shopping.

居住点：在小区出入口布点，缓解公交"最后一公里"及短途出行的问题。

Residential areas–distribute the public bicycles at the entrance and exit in neighborhood to release the "last mile" traveling problem.

公建点：设置在商业区的公共自行车服务点
Public constructions-set public bicycle service point in business area.

居民点：设置在居住区的公共自行车服务点
Residential area-set public bicycle service point in residential area.

珠江新城集国际金融、贸易、文娱、行政和居住等城市功能设施于一体，应完善非机动车道路设施、增强区域的可达性，方便行人和自行车的出行，吸引更多的市民到珠江新城来休闲。广州在 BRT 走廊沿线已有 5000 辆公共自行车，系统运营良好，建议在珠江新城设置公共自行车系统，并纳入现有的公共自行车系统当中，实现各个服务点的通租通还。

Zhujiang New Town has concentrated international finance, trade, business, entertainment, administration and housing and other urban functions and facilities. It has improved non-motorized traffic facilities, enhanced accessibility of the region, facilitated pedestrian and bicycle travel, to attract more people to the Zhujiang New Town. Guangzhou has 5,000 public bicycles along the BRT corridor, the system operates in good condition. We suggest constructing public bicycle system in Zhujiang New Town, and incorporate it in the current bicycle system to achieve connected renting and returning in every service point.

在珠江新城内公交站点、地铁站点、绿道及共建点、城市兴趣点、居住区设置公共自行车服务点，方便市民到达珠江新城，骑行娱乐。

The bus stations, subway stations, greenway and public constructions, city's interesting sites, and public bicycle service points in residential areas facilitate citizens reaching Zhujiang New Town and riding and entertaining in this area. The image below is the public bicycle system in Zhujiang New Town.

珠江新城公共自行车系统布点图 *Distribution graph of public bicycle systems in Zhujiang New Town*

Chapter Three

兰州绿道改善建议
Lanzhou Greenway Improvement Suggestion

参考方法
Reference Method

绿道应保持其自身的连续性,特别是在复杂的交通条件下,更是需要保证其整体连续,方便绿道使用者,安全地通过交通复杂的地段。绿道的连通可根据不同的现状情况做出具有适应性的设计,如在天桥处绿道中断,可设置有颜色的混行车道连接站点两侧车道。而对于交叉口,则可根据现状交叉口设计,设置分离的自行车通道。

Continuity is essential for greenway, especially when it is in complicated transportation condition, because continuity helps greenway users get through complicated traffic areas conveniently and safely. The connectivity of greenways can be designed in different ways based on different road status. For example, when the greenway is interrupted by the footbridge, a painted cycle lane mixed with vehicle lane can connect the separated sections. As for the intersection, dedicated bike crossing can be designed according to the current status and signal phasing of the intersection.

黄河风情线绿道现状。

The greenway status in yellow river scenery line.

实践案例
Practice

黄河风情带在排洪口处中断。
The Yellow River scenery belt is interrupted by flood drainage.

黄河风情带被桥梁阻隔。
The Yellow River scenery belt is obstructed by bridge.

黄河风情带在排水泵站处中断。
The Yellow River scenery belt is interrupted by drainage station.

Chapter Three

滨河路 Binhe Road

现状 Currents Status

现状风情线公共空间宽度介于 5~6 米之间，有较多市民在公共空间上游玩和休憩。道路外侧有 1.5 米宽自行车道、3 米宽的绿化带和人行道。

The public space in the current scenery line is about 5-6 meters wide, and there are many people playing and rest in the public space. The bicycle lane in the outer side of road is 1.5meters wide and the green belt and sidewalks are 3meters wide.

南滨河路(中山桥-白云观段)
Nanbinhe Road (Zhongshan Bridge-Baiyunguan Section)

现状照片
Current condition

改善 Improvement

抬升现有自行车道地面，保证行人、自行车、非机动车的有效分离。考虑现状风情线公共空间游人较多，不在风情线内部设计自行车道。

Enhancing the current bicycle lane, keep the pedestrian, bicycle and non-motor vehicles effectively separated. As for the large amount of travelers in the public space of scenery line, thus the bicycle line inside the scenery line is not necessary.

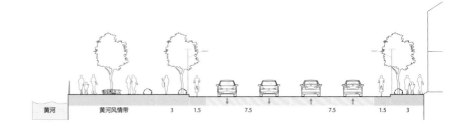

现状断面
Current road section

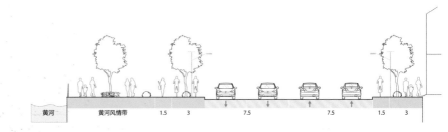

改善断面
Proposed road section

实践案例
Practice

南滨河路（雷坛河-小西湖黄河桥）

Nanbinhe Road (Leitan River-Xiao Xihu Huanghe Bridge)

现状照片
Current condition

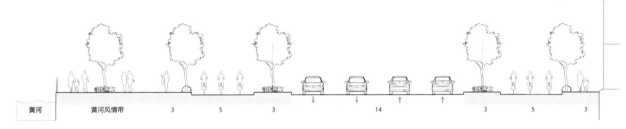

现状断面
Current road section

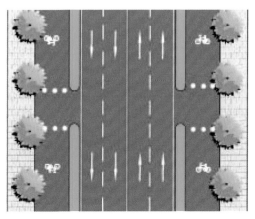

改善平面
Proposed plan

Chapter Three

● **南滨河路(石炭子桥–银滩大桥)**

Nanbinhe Road (Shitanzi Bridge- Yintan Bridge)

现状照片
Current condition

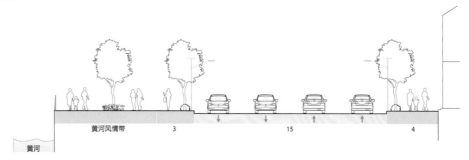

现状断面
Current road section

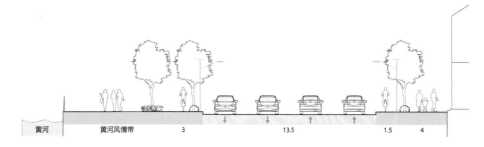

改善断面
Proposed road section

实践案例
Practice

北滨河路（西沙大桥-七里河桥）

Beibinhe Road (Xisha Bridge-Qilihe Bridge)

现状照片
Current condition

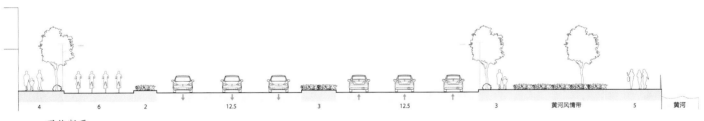

现状断面
Current road section

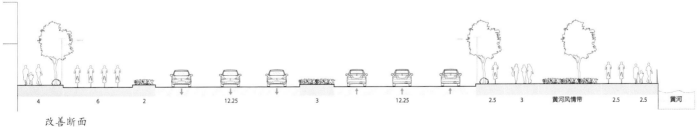

改善断面
Proposed road section

Chapter Three

北滨河路（七里河桥–消防大队）
Beibinhe Road (Qilihe Bridge-Fire Brigade)

现状照片
Current condition

现状断面
Current road section

改善断面
Proposed road section

北滨河路（消防大队-欧洲阳光城）

Beibinhe Road (Fire Brigade-Europe Sunshine City)

现状照片
Current condition

现状断面
Current road section

改善断面
Proposed road section

Chapter Three

北滨河路（欧洲阳光城-雁滩大桥）
Beibinhe Road (Europe Sunshine City- Yantan Bridge)

现状照片
Current condition

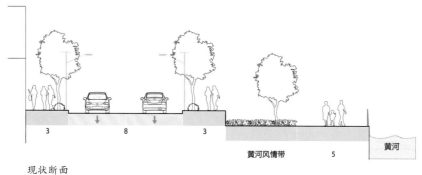

现状断面
Current road section

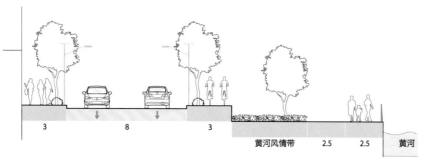

改善断面
Proposed road section

南滨河路（雁滩大桥-雁宁路）

Nanbinhe Road (Yantan Bridge-Yanning Road)

现状照片
Current condition

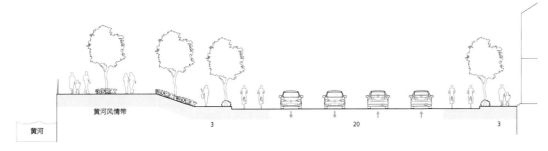

现状断面
Current road section

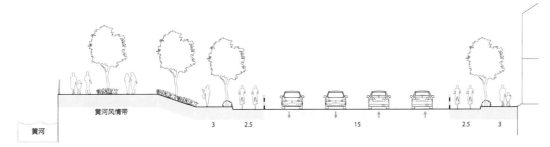

改善断面
Proposed road section

Chapter Three

南滨河路（雁宁路–水车园）
Nanbinhe Road (Yanning Road-Shuicheyuan)

现状照片
Current condition

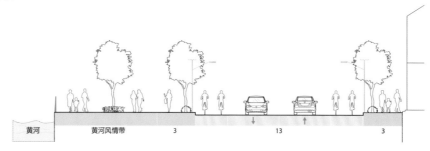

现状断面
Current road section

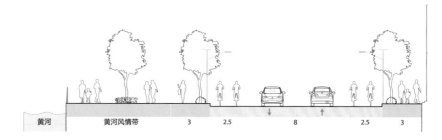

改善断面
Proposed road section

实践案例
Practice

南滨河路（雁滩大桥-雁宁路）
Nanbinhe Road (Shuiche Park-Zhongshan Bridge)

现状照片
Current condition

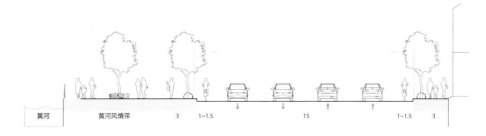

现状断面
Current road section

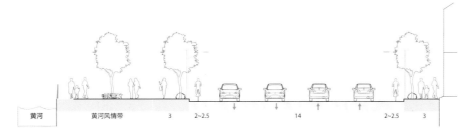

改善断面
Proposed road section

177

Chapter Three

广州淘金 – 建设新村改造建议
Taojin-Jianshe Xincun Reform Suggestion

 项目区位

Project Location

项目位于广州市越秀区环市东商业圈，号称"广州尖东"，是广州市旧城核心区，片区相关设施发展较为完善。

The Project is located in Guangzhou Huanshidong business area in Yuexiu District. It is called "Tsim Sha Tsui East of Guangzhou ". It is the core area of Guangzhou old city. The Facilities in this area are quite completed.

建设新村位于广州市环市东商贸圈，周边酒店、商业、办公楼密集，人流密集，是体现广州市容市貌的重要社区。而淘金片区素有环市东"后花园"的美誉，不但是一个成熟的居住社区，也是发达的金融、商务与文化产业区。

Jianshexincun is located in Guangzhou Huanshidong trading and business area. The nearby hotels, business, office buildings and population are very dense. It is an important area of manifesting the image of Guangzhou city. Taojin Road enjoys a good reputation as the "Back Garden" of Huangshidong. It is not only a popular residential area, but also a prosperous financial and business industry area.

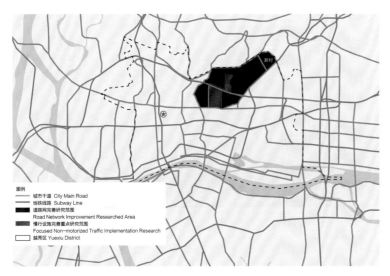

淘金 – 建设新村片区区位图
The Geographical Map of Taojin-Jianshexincun

2 道路网络分析及建议
Road Network Analysis and Suggestion

2.1 道路网结构
Road Network Structure

①建设新村和淘金片区属于成熟片区，路网较为完善，支路网密度较高。
②建设新村和淘金片区存在较多社区内街，多设有铁门，但白天处于敞开状态；淘金片区多处存在高差，多处设有人行台阶，步行网络较为便捷。
③建设五马路往环市东路现状有内街通行，建议打通围墙。
④淘金坑路北段现状被围蔽为停车场，建议开放道路。

① The construction of Jianshexincun and Taojin area belong to developed area with completed road network and dense road branches.
② Jianshexincun and Taojin area have many streets inside neighborhoods with iron gates, but these gates are often open in the daytime. Height differences exist in many places in Taojin area, and there are a lot of stairs, though the walking network is relatively convenient.
③ There are streets inside Jianshewu Road and Huangshidong Road, and it is recommended to break the wall for the consistency of passage.
④ The north section of Taojinkeng Road is enclosed as a parking lot. We suggest opening up as a passage.

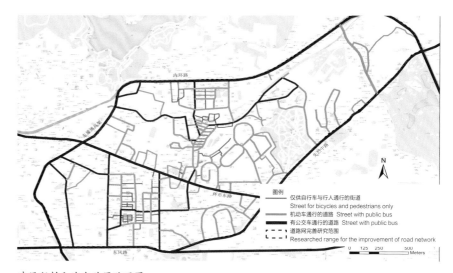

建设新村和淘金片区路网图
The road network in Jianshexincun and Taojin area

建设五马路内巷打通围墙。
Open up the walls for passage in Jianshewu Road.

淘金坑路围蔽路段。
The enclosed road section in Taojinkeng Road.

2.2 道路流量分析
Road Traffic Analysis

根据现状调查流量分析可知：
①主要机动车交通流量集中在外围干道，片区内部道路机动车交通量较小，片区整体运行情况良好。外围干道以交通功能为主，应保障机动车路权。内部道路以慢行交通为主，应保障行人路权，提高慢行安全性和舒适性。
②片区自行车交通流量不大，主要集中在环市东路、华乐路、淘金路、建设大马路、淘金东路等。由于片区道路空间有限，内部道路主要保障行人空间，除自行车流量较大的道路外，原则上不设独立的自行车道。片区内部道路可设置多种减速措施，在自行车与机动车混行的情况下保障自行车的安全。

According to the current situation, we can conclude the result that:
① The current motor vehicles are mainly concentrated in the main road outside, and the traffic flow is small within the area. The traffic flow in this whole area is in a good condition. The outer side of main road is mostly functioned as transportation. The motor vehicles passing should be guaranteed. The inner road is mostly functioned as non-motorized traffic, and pedestrians passing should be guaranteed to enhance the safety and comfortable experience.
② The bicycle flow in this area is small, mainly concentrated in Huanshidong Road, Huale Road, Taojin Road, Jianshe Avenue and Taojindong Road etc. The space is limited in this area, and the inner road ensures pedestrians' space. Except the road with large bicycle flow, bicycle lane is in principle not independent. There are various kinds of speed bumps in the inner road of this area to protect the safety of bicycles when mixing with motor vehicles.

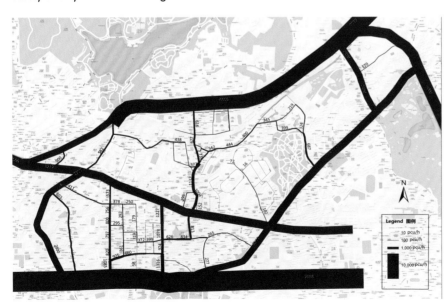

现状晚高峰机动车流量

Current motor vehicle traffic flow in evening peak

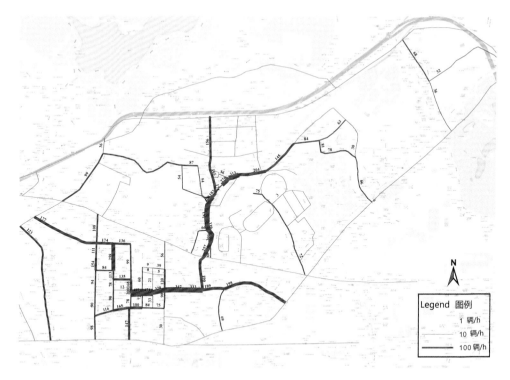

现状晚高峰自行车流量
Current bicycle traffic flow in evening peak

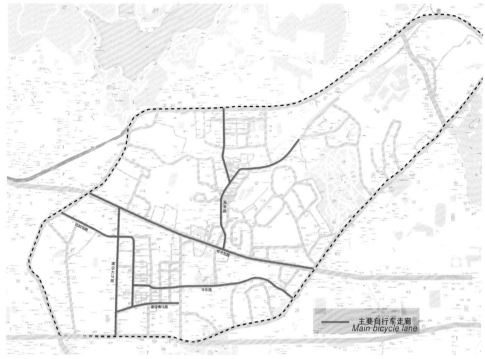

自行车走廊建议
Bicycle corridor proposal

2.3 道路交通组织优化
Road Traffic Organization Optimization

建设新村和淘金片区为成熟的居住片区，应坚持内外分离、屏蔽过境交通的原则，优化建设新村和淘金片区交通组织。为居民提供一个舒适、安全的居住环境。

Jianshexincun and Taojin area are mature residential areas, and they adhere to the principle of internal and external separated for keeping distance with transportation and optimizing the construction of transportation within the area. Therefore, provide a comfortable and safe environment for residents.

①建设大马路调整为双向通行。建设横马路为建设新村内部道路，承担了大量东风路经建设大马路－建设横马路－建设六马路往环市东路的车流。建设大马路调整为双向通行后，建设横马路过境功能削弱，沿线慢行条件得到提升。
②建设二马路东五街调整为双向通行，建设二马路东四街调整为步行道。建设二马路东五街和建设二马路东四街现状均为单行道，但由于靠近建设新村菜市场，占道经营行为较多，秩序较为混乱，车流量较少。建设二马路东五街调整为双向通行，南侧道路调整为步行道，整顿占道经营现象，优化流线组织，改善慢行环境。
③建设三马路南段（华乐路－建设横马路）调整为步行道。
④建设横马路（建设三马路－建设四马路）调整为步行道。
⑤淘金路交通组织保持现状不变。淘金片区路网呈鱼骨架结构，以淘金路为主轴，向外辐射。淘金路主要承担对外交通功能，北接恒福路和内环路，南接华乐

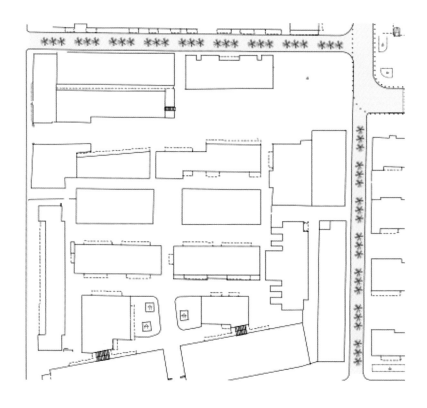

步行道调整示意图　Walking lane adjustment

路，连接广州干线路系统和建设新村、华乐片区。内部道路主要承担疏散功能。

① Jianshe Avenue is a two-way traffic road, and Jianshe road is a road inside Jianshexincun, which bears a large number of the traffic flow that passes from Jianshe Avenue to Jiansheer Road and Jiansheliu Road. After the opening of Jianshe Avenue's two-way traffic, the crossing from Jiansheheng Road is weakened, and the non-motorized traffic lines along the road are improved.

② After the opening of Jianshe Avenue's two-way traffic, Dongsi Street in Jiansheer Road has been adjusted to pedestrian lane. Dongwu Street in Jiansheer Road and Dongsi Street in Jiansher Road are one way lane, however, as they close to the food market in Jianshexincun and many vendors occupation on the roads, thus have less traffic flow. Dongwu Street in Jiansheer Road is adjusted to two-way lane and the south side of road is adjusted to walking lane. The vendor occupation has been removed and organizational flow has been improved. The non-motorized traffic environment is enhanced.

③ The south section of Jianshesan Road (Huale Road–Jiansheheng Road) is adjusted to pedestrian lane.

④ Jiansheheng Road (Jianshesan Road–Jianshesi Road) is adjusted to pedestrian lane.

⑤ The transportation organization in Taojin Road will stay in the current status. The road network in Toajin Area is presented as a fish bone structure. Taojin Road is the main road and radiates branches outward. Taojin Road is mainly responsible for the traffic function. North extends to Hengfu Road and Neihuan Road, and the south stretches to Huale Road, which connect the main road system in Jianshexincun and Huale area in Guangzhou. The inner road is mainly responsible for evacuating the traffic.

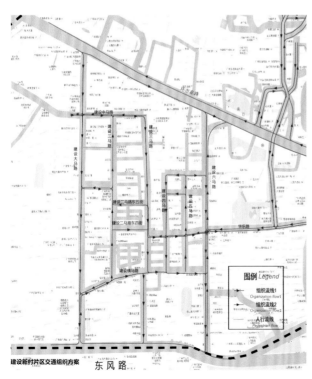

Organization scheme of the traffic in Jianshexincun area.

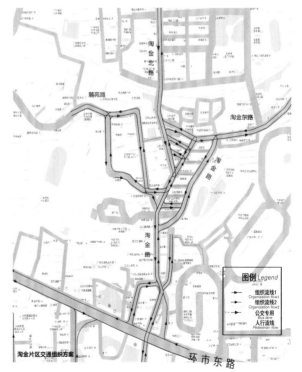

Organization scheme of the traffic in Taojin area.

Chapter Three

3 道路平面及横断面设计
Road Plan and Cross Section Design

3.1 设计要点
Design Key Point

3.1.1. 交通组织优化
Traffic Organization Optimization

在道路规划阶段，明确道路空间的功能分区，避免或减少机动车流与慢行交通的相互影响。

In the stage of road planning, road space should be divided clearly for different functions, avoiding or reduceing the conflict between vehicles and non-motorized traffic.

分隔机动车与自行车道，除了用传统的护柱等设施外，还可结合利用路边停车和公共自行车系统。（巴塞罗那）

Separating motor vehicles and bicycle lane, except application of traditional bollard and other facilities, can also combine the roadside parking and public bicycle system. (Barcelona)

实践案例
Practice

3.1.2 拓宽人行道
Expand the Sidewalk

通过人行道向车行道扩张，增加慢行交通系统空间，减少机动车流量，同时也减少行人、非机动车暴露在机动车流中的时间。

Expanding the sidewalk to the outer area can increase more space for non-motor traffic system, at the same time reduce the time of pedestrian and non-motor vehicles exposuring in the motor vehicles' flow.

在没有转弯车流的角落，人行道可做成直角。纽约曼哈顿联合广场改造，扩大了行人的空间，收缩交叉口面积。改造后交通事故伤亡降低了26%，74%的使用者满意现在的改变。

In the corners of no turning flow, sidewalk can be made as a right angle. The transformation in New York Union Square, Manhattan has expanded the pedestrian space and narrowed the area of intersection. The traffic accident have casualties decreased by 26% after the transformation, and 74% users are satisfied with the change.

Chapter Three

3.1.3 车行道宽设为 3.25~3.5 米
Set Width of the Traffic Lane to 3.25~3.5m

合理设置机动车道宽度，根据车流量及通过车辆类型，选择 3.25 米或 3.5 米的标准车道，避免车道过宽侵占慢行空间。

Set reasonable width to the traffic lane, according to vehicle's flow and the type of vehicles to set standard traffic lane within the width of 3.25m to 3.5m, to avoid the width of traffic lane exceeding the non-motor vehicles lane.

3.1.4 抬升街道
Uplifted Street Lane

抬升机动车道，使之与人行道平齐，同时在保证足够的限制性宽度下用护柱加以隔离，使机动车提前降低车速，构建行人与机动车宁静和谐的出行环境。

Uplifting motor vehicles lane can make it to the same height with the sidewalk. At the same time, give enough restrictions to the width of the bollards for separating and reducing the speed of motor vehicles. Therefore, construct harmonious environment for pedestrian and motor vehicle.

抬升街道，采用人行道铺装。
Uplifted street lane, pave for the sidewalk.

3.1.5 增设护柱以保障人行空间
Add Bollards to Ensure Sidewalk Space

规范机动车出入口设计，通过工程手段，尽量减少机动车出入口对慢行交通系统的不良影响；同时，结合抬升街道地面的措施防止机动车对人行空间的占用。通过增设人行道护柱来弥补交通管理的不足，规范驾驶员的行为。

Regulate the entrance of motor vehicles through the means of engineering to reduce effects of motor vehicles' entrance on non-motor traffic. Meanwhile, combine with the lifted street to prevent motor vehicles' occupation on sidewalk. By setting up the bollard on sidewalk to replace the shortage of traffic management can regulate the behavior of drivers.

2002 年 - 2009 年，法国巴黎投入了 1500 万欧元在全市设置了 33.5 万个人行道护柱，禁止机动车占用人行道。

From 2002 to 2009, Paris spent 15 million Euros on 335 thousand bollards in the whole city to prevent motor vehicle's occupation on sidewalk.

将机动车出入口道路抬高至与人行道平齐，并铺设人行铺装，保持人行道连续。（伦敦）

Raise vehicle access to the same hight as the sidewalk and use pedestrian pavement to keep the continuity of the sidewalk. (London)

3.1.6 设计弯曲的道路
Design of Curved Road

通过设计弯曲的道路来降低机动车速度，提高街道安全性。道路两侧可以交错拓宽人行道、加设停车位、花坛、座椅等设施来达到弯曲道路的设计。

Curved road can reduce vehicle speed and improve road safety. Broadening one side of the sidewalk on one section and the other side on another section in concessive in turn with parking space, planting and seats can make a curved road.

弯曲的道路设计，两侧可加宽人行道等。
The curved road design helps to broaden both sides of the sidewalk.

3.1.7 增加座椅
Add More Seats

路侧及其他慢行空间适当的增加座椅，以供行人休憩，同时有助于形成宁静舒适的慢行空间，增加区内的生活气息。

On the road side or other non-motor traffic areas, one can appropriately add seats for pedestrians to take rest, and helps to form a tranquil and comfortable space which will brings more life flavor.

错落有致的路侧座椅。（布达佩斯）
The appropriately displayed chairs.(Budapest)

3.2 道路平面及横断面设计
Road Plane and Cross Section Design

3.2.1 设计依据
Design Basis

①本次设计基于不拆迁/减少投资的原则，在道路红线范围内对路权重新分配，优化道路空间，保障行人通行的舒适与安全性。②在现状道路设施和道路流量调查基础上，对设施进行优化。③由于道路空间有限，人行道宽度不足，所以淘金-建设新村片区除少数几条自行车流量较大的道路外，原则上不设独立的自行车道。片区内部道路可通过设置多种减速措施，降低片区内车速，保障自行车的安全性。

① The design is based on principle of no demolition and the reduction of investment, to reallocate the area within the red line for optimizing the road space and ensure the comfortable feeling and safety of pedestrian traffic.
② On the basis of current situation of road facilities and road traffic investigation, optimize the facilities.
③ Due to the limited road space and insufficient sidewalk width, in Taojin and Jianshexincun area, it is not necessary to set dedicated bikelane on all streets, except for a few main roads with larger bike flows. Reducing the traffic speed through traffic calming designs will ensure the safety of cycling.

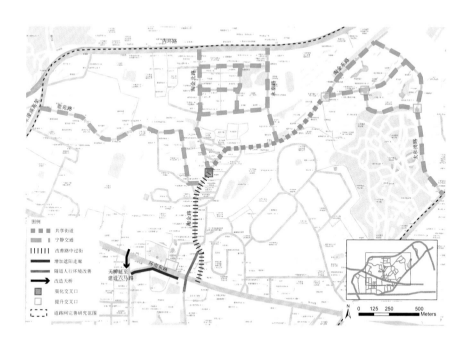

淘金道路　　Taojin Road
平面设计总图　　Planedesign general layout

3.2.2 改善措施
Improvement Measures

①拓宽人行道，保障行人空间。
②调整交通组织，部分路段设置为步行街。
③抬升路面，采用人行铺装，设置为共享街道。
④改善路中过街，增设路中过街安全岛。

① Broaden sidewalk to ensure the pedestrian space.
② Adjust the traffic organization and set up some sections of the road as pedestrian street.
③ Uplift the sidewalk, pave for sidewalk and set it as a shared public street.
④ Improve the road crossing, and set safety island in the middle of the pedestrian crossing.

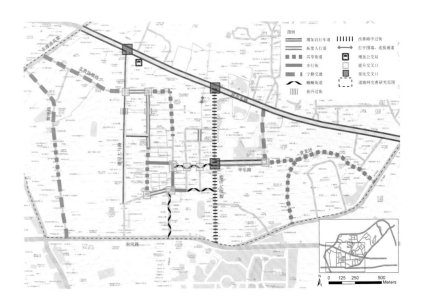

建设新村道路平面设计总图
The roads of Jianshexincun plane design general layout.

3.2.3 建设大马路
Jianshe Avenue

现状 Currents Situation

①现状建设大马路（环市东路－建设横马路）段为北往南单向通行，建设大马路（建设横马路－东风路）段为双向通行。全段均为两个车道，环市东路－建设横马路段设置路了内停车。
②据调查，建设大马路（环市东路－建设横马路）段晚高峰每小时标准车当量约为800，建设大马路（建设横马路－东风路）段晚高峰每小时标准车当量约为1100。自行车流量约100辆/小时，自行车流量较大。
③全段人行道有效宽度均大于2米，人行设施条件较好。

① The current Jianshe Avenue (Huanshidong Road–Jiansheheng Road) section from north reach to south is one way traffic, and Jianshe Avenue (Jiansheheng Road–Dongfeng Road) section is two- way traffic. The whole section is divided into two, in which constructed roadside parking in Huangshidong–Jiansheheng Road section.
② According to the survey, the motor vehicles' flow in evening peak in Jianshe Avenue (Huanshidong Road–Jiansheheng Road) section is 800pcu/h, while Jianshe Avenue (Jiansheheng– Dongfeng Road) section is 1100 pcu/h .
③ The width of the whole section of sidewalk is wider than 2m, and the sidewalk facilities are in good condition.

设计方案 Design Plan

建设大马路调整为双向通行，保留现有路内停车位，沿线交叉口和主要出入口设置人行道护柱，防止机动车停车占用行人空间。

Jianshe Avenue is adjusted to two-way traffic, and retains the roadside parking area. The intersections along the road and the main entrance are equipped with bollard to prevent motor-vehicles occupation on pedestrian space.

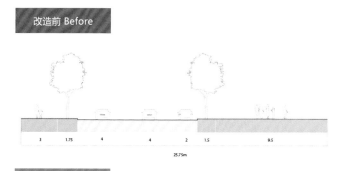
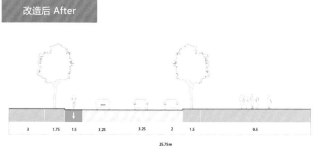

建设大马路（环市东路-建设横马路）断面
Jianshe Avenue (Huangshidong Road-Jiansheheng Road) Section

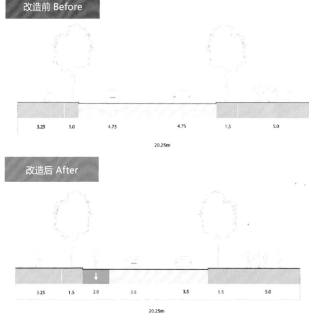

建设大马路（建设横马路-东风路）断面
Jianshe Avenue (Jiansheheng Road-Dongfeng Road) Section

Chapter Three

环市东路 / Huanshidong Road 建设中马路 / Jianshezhong Road 建设横马路 / Jiansheheng Road

建设大马路平面方案
Plane plan of Jianshe Avenue

3.2.4 建设二马路
Jiansheer Road

现状 Current Situation

①建设二马路现状为南往北单行道，单向两车道，且车道宽度均大于 3.5m。道路设有路内停车，且违章停车现状较为普遍。
②建设新村菜市场门口占道摆摊现象较为严重，且较多行人随意穿行马路，秩序混乱。
③据调查，建设二马路晚高峰每小时标准车当量约为 350。车行道空间未得到有效利用。
④部分路段人行道有效宽度小于 2 米，人行设施有待完善。

① The current Jiansheer Road is one-way traffic from south to north, for two lanes and the width of lanes is more than 3.5m. There are roadside parking in the lane, but illegal parking is quite common.
② The vendor occupation on the food market of Jianshexincun is relatively common, and many pedestrians cross the road randomly. The road condition is in a mess.
③ According to the survey, the traffic flow in evening peak hours in Jiansheer Road is 350pcu/h. The traffic space has not been properly used.
④ The effective width of the pavement in some sidewalk sections is less than 2m, and the facilities need to be improved.

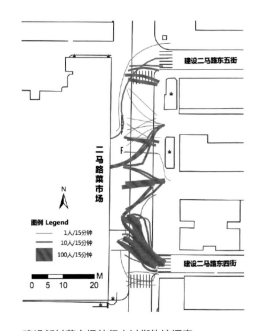

建设新村菜市场外行人过街轨迹调查
Pedestrian trajetory survey outside Jianshexincun food market.

设计方案 Design Plan

建设二马路北段（建设中马路－广州市规划勘测设计研究院北侧道路）车行道宽度调整为 3.5 米，释放人行空间。

The width of north section of Jiansheer Road (Jianshezhong Road to Guangzhou City Planning Survey and Design Research Institute on the north side of the road) is adjusted to 3.5m to release more sidewalk space.

建设二马路中段（广州市规划勘测设计研究院北侧道路－建设二马路东五街）抬升路面，铺设人行道铺装，设置护柱分隔，保证建设新村菜市场门口行人过街安全性。

The road in the middle of Jiansheer (Guangzhou City Planning Survey and Design Research Institute to Dongwu Road in Jiansheer Road) is lifted; sidewalk is paved and set bollards for separation to ensure the safety of the road crossing of the food market in Jianshexincun.

建设二马路南段（建设二马路东五街－建设横马路）抬升路面，铺设人行道铺装，设置护柱分隔，车行道宽度调整为 3.25 米，设置停车位和卸货泊位。

The road in south section of Jiansheer Road (Dongwu Street in Jianshe Road to Jiansheheng Road) is lifted, sidewalk is paved and set bollard for separation. The width of traffic lane is adjusted to 3.25m, parking places and areas set for loading and uploading goods.

改造前
Before renovation

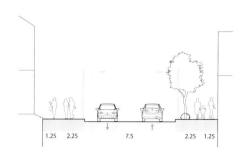

改造后
After renovation

建设二马路（建设中马路－规划院）断面
The road section of Jiansheer Road

建设二马路平面方案 *The Plane of Jiansheer Road*

Chapter Three

现状
Current situation

建议
Proposal

3.2.5 建设三马路
Jianshesan Road

现状 Current Situation

①建设三马路（建设中马路－建设二马路东五街）段路人行道较窄。大部分路段人行道宽度仅1米。
②建设三马路－建设横马路交叉口为畸形交叉口，建设三马路北往南车流只能经建设横马路进行转换，不能经建设三马路到达东风路。现状存在较多违章行驶。

① The sidewalk in Jianshesan Road (Jianshezhong Road–Dongwu Street in Jiansheer Road) section is narrow and most of the sidewalk in this section is only wide 1 m.
② The intersection of Jianshesan Road–Jiansheheng Road is twisted. The traffic flow from north to south can only be transfered in Jiansheheng Road, but cannot reach Dongfeng Road through Jianshesan Road. There are many illegal driving exists.

改善案例
Reference

实践案例
Practice

设计方案 Design Plan

建设三马路北段（建设中马路 – 建设二马路东五街）抬升路面，采用人行道铺装，设置护柱，与建设中马路连接。

North of Jianshesan Road (Jianshezhong Road to Dongwu Street in Jiansheer Road) is lifted, sidewalk is paved and set bollard to connect with Jianshezhong Road.

建设三马路中段（建设二马路东五街 – 华乐路）保持不变。

The middle of Jianshesan Road (Dongwu Street in Jiansheer Road to Huale Road) will remain untouched.

建设三马路南段（华乐路 – 建设横马路）调整为步行道，抬升路面，采用人行道铺装。

South of Jianshesan Road (Huale Road–Jiansheheng Road) is adjusted to pedestrian lane, and road will be lifted and paved for sidewalk.

建设三马路（建设横马路 – 东风路）调整为北往南单行单车道，设置为弯曲的道路，降低车速，释放人行空间。

Jianshesan Road (Jiansheheng Road–Dongfeng Road) is adjusted to one-way lane from north to south. Set curved lane to lower the speed of vehicles and release more pedestrian space.

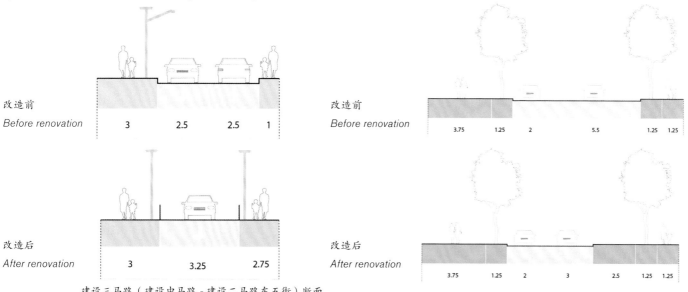

改造前
Before renovation

改造后
After renovation

建设三马路（建设中马路 - 建设二马路东五街）断面
The road section in Jianshesan Road (Jianshezhong Road to Dongwu Street in Jiansheer Road)

建设三马路（建设横马路 - 东风路）断面
Jianshesan Road (Jiansheheng Road-Dongfeng Road) section

Chapter Three

建设三马路平面方案 Jianshesan Road Plane Scheme

建设三马路 - 华乐路交叉口现状

Intersection of Jianshesan Road and Huale Road Current Situation

建设三马路 - 华乐路交叉口建议

Intersection of Jianshesan Road and Huale Road Proposal

3.2.6 建设中马路

Jianzhezhong Road

现状 Current Situation

①建设中马路（建设大马路－建设二马路）段现状路面较宽，为双向三车道。据调查，现状晚高峰每小时标准车当量约为370，道路空间未得到有效利用。

②建设三马路（建设二马路－建设三马路）段现状有建设大马路小学出入口，上下学有较多小学生出入，存在交通安全隐患。

① The road in Jianzhezhong Road (Jianshe Avenue to Jiansheer Road) is wide, and it is a three-way lane. According to the survey, the current motor vehicles flow in evening peak is about 370pcu/h. The road space has not been effectively utilized.
② Jianshesan Road(Jiansheer Road to Jianshesan Road) section has entrance and exit of the Primary School of Jianshe Avenue, and there are many pupils coming in and out of the school, which causes security problems

设计方案 Design Plan

建设中马路西段（建设大马路－建设二马路）调整为双向双车道，释放人行空间。

Adjust the west section of Jianshezhong Road (Jianshe Avenue to Jiansheer Road) to two-way lane to release more sidewalk space.

建设中马路东段（建设二马路－建设三马路）车行道宽度调整为3.5米，释放人行空间。同时抬升路面，采用人行道铺装，设置护柱。

In the east side of Jianshezhong Road (Jiansheer Road to Jianshesan Road), the width of traffic lane can be adjusted to 3.5m to release more sidewalk space. Uplift the sidewalk, pave the road surface and set bollards.

改造前
Before renovation
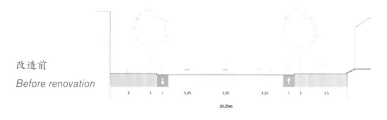

改造前
Before renovation
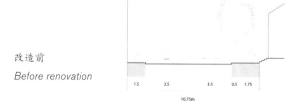

改造后
After renovation
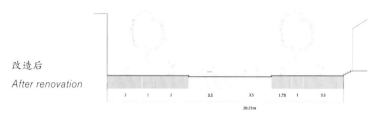

改造后
After renovation
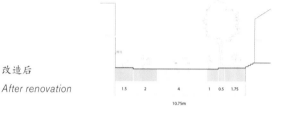

建设中马路（建设大马路 - 建设二马路）断面
Jianshezhong Road (Jianshe Avenue to Jiansheer Road) section

建设中马路（建设二马路 - 建设三马路）断面
Jianshezhong Road (Jiansheer Road to Jianshesan Road) section

Chapter Three

建设中马路平面方案 Jianshezhong Road Plane Scheme

3.2.7 建设横马路

Jianzheheng Road

现状　Current Situation

①建设横马路（建设大马路－建设三马路）段现状为单向两车道，建设横马路（建设三马路－建设六马路）段为单车道，并设有路内停车。由于现状建设大马路（建设横马路－环市东路）段为北往南单行，东风路往环市东路车流较多通过建设横马路绕行，建设横马路承担较多过境交通。
②据调查，建设横马路（建设大马路－建设二马路）晚高峰每小时标准车当量约为900，建设横马路（建设二马路－建设三马路）晚高峰每小时每小时标准车当量约为650，建设横马路（建设三马路－建设六马路）晚高峰每小时标准车当量约为400。
③建设横马路（建设三马路－建设六马路）段有建设六马路小学，上下学有较多小学生出入，存在交通安全隐患。

① One section of Jiansheheng Road (Jianshe Avenue to Jianshesan Road) two lanes of one way. One section of Jiansheheng Road (Jianshesan Road to Jiansheliu Road) has one lane as well as on-street parking. Because one section of Jianshe Avenue (Jiansheheng Road to Huanshidong Road) is a one-way lane from north to south, so that there are large traffic flows from Dongfeng Road to Huanshidong Road, and many vehicles make a detour on Jiansheheng Road which brings a large through traffic flow on Jiansheheng Road.
② According to the survey, the traffic flow in peak hours of Jiansheheng Road (Jiansheer Road to Jianshesan Road) is 900 pcu/h, and the traffic flow in the

建设横马路 - 建设三马路路口为畸形交叉口，建议于交叉口处封住建设横马路，避免西往东车流穿行，净化建设横马路慢行环境。

The intersection from Jiansheheng Road to Jianshesan Road is twisted, and it seals the intersection in Jiansheheng Road to avoid the traffic passing from west to east and clean the non-motorized traffic environment of Jiansheheng Road.

some perial of Jiansheheng Road(Jiansheer to Jianshesan Road) is 650 pcu/h. The evening peak of Jiansheheng Road (Jianshesan Road to Jiansheliu Road) is 400pcu/h.

③ There are many pupils comming in and out of the Primary School of Jianshe Avenue,in Jiansheheng Road (Jianshesan Road to Jiansheliu Road) section, which causes security problems.

设计方案 Design Plan

建设横马路西段（建设大马路 – 建设二马路）保持单向双车道不变。

The middle section of Jiansheheng Road (Jianshe Avenue to Jiansheer Road) is adjusted to one-way lane from west to east.

建设横马路中段（建设二马路 – 建设三马路）调整为西往东单行。

The east side of Jiansheheng Road (Jiansheer Road to Jianshesan Road) is adjusted to lifted sidewalk.

建设横马路东段（建设三马路 – 建设四马路）调整为步行道，抬升路面。

建设横马路（建设四马路 – 建设六马路）抬升路面，采用人行铺装。全段设置护柱，防止机动车停车占用行人空间。

The sidewalk in Jiansheheng Road (Jianshesi Road to Jiansheliu Road) can be lifted and paved. The whole section will be set bollards for prevention of motor vehicles' occupation on sidewalks.

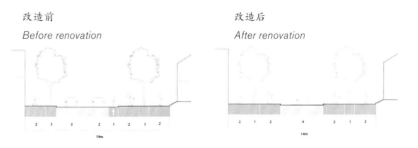

改造前 Before renovation　　改造后 After renovation

建设横马路（建设二马路 - 建设三马路）断面
Jiansheheng Road (Jiansheer Road to Jianshesan Road) section

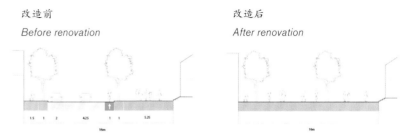

改造前 Before renovation　　改造后 After renovation

建设横马路（建设三马路 - 建设四马路）断面
Jiansheheng Road (Jianshesan Road-Jianshesi Road) section

Chapter Three

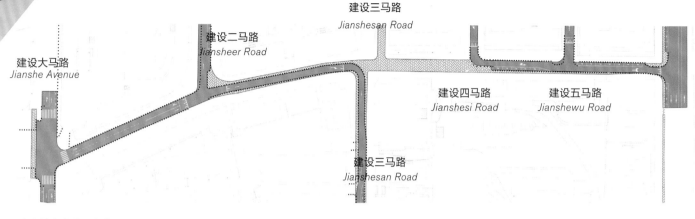

建设横马路平面方案 Jiansheheng Road Plane Scheme

3.2.8 建设二马路东四街、建设二马路东五街
Dongsi Street in Jiansheer Road, Dongwu Street in Jiansheer Road

现状 Current Situation

①建设二马路东五街和建设二马路东四街临近建设新村菜市场，现状均为单行道。但由于较多小商贩占道经营，现状通行能力较差，车流量较小。
②建设二马路东四街现状有建设新村–淘金片区环线公交车经过，人行道较窄。

① Dongwu Street and Dongsi Street near the food market are all one-way lanes currently. The vendors accupation has weakened the traffic capacity and the vehicles' flow is small.
② Dongsi Street in Jiansheer Road currently has buses pass by from Jianshexincun to Taojin area. The sidewalk is narrow.

建设二马路东四街抬升路面，采用人行铺装，规范占道摆摊行为，提升品质。（左图可供参考）

The road on Dongsi Street in Jiansheer Road can be raised with sidewalk pavement. The vendors' occupation should be managed to improve the space quality. (Left photo is an example.)

实践案例 Practice

设计方案 Design Plan

建设二马路东五街调整为双向通行，南向平行道路调整为步行道，抬升路面，采用人行铺装。原南向平行道路上公交线路调整至建设二马路东五街。

Dongwu Street in Jiansheer Road is adjusted to two-way traffic lane, and the south parallel road can be adjusted to pedestrian street. The road can be lifted and paved. The bus line in original south parallel road can be moved to Dongwu Street in Jiansheer Road.

建设二马路东五街断面
Dongsi Street in Jiansheer Road

建设二马路东四街、建设二马路东五街平面
Dongsi Street in Jiansheer Road; Dongwu Street plane in Jiansheer Road.

3.2.9 华乐路
Huale Road

现状 Current Situation

①华乐路为联系建设新村片区与华乐片区、淘金片区主要的通道。华乐路现状为双车道，但车道较宽，为 4.85 米。
②据调查，华乐路（建设六马路-淘金路）晚高峰每小时标准车当量约为 800。自行车流量约为每小时 350 辆，自行车流量较大。

Chapter Three

取消左侧垂直道路停车，调整为路内平行道路划线停车，拓宽人行道空间。
Cancel the vertical road parking on the left side, and adjust to parallel roadside parking to release more pedestrian space.

设置隔离带，分隔自行车道与机动车道，保障自行车的安全
Set separating belt to isolate bicycle lane and vehicles' lane, to ensure the safety of bicycles.

利用现状道路红线形成弯曲的道路，降低车速
Use the current red line to form a curved road for decorating vehicles' speed.

① Huale Road is the connection of Jianshexincun area, Huale area and Taojin area. In current, Huale Road is a wide two-way traffic lane, and the width is 4.85 meters.

② According to the survey, the traffic flow in evening peak in Huale Road (Jiansheliu Road to Taojin Road) is about 800pcu/h. The bicycle flow is about 350/h, and the amount of bicycles is large.

设计方案 Design Plan

压缩现有车行道，保证车行道宽度为 3.25~3.5 米，释放人行空间。

Narrow the existing roadway, and ensure the width of roadway on 3.25-3.5m to release more pedestrian space.

在现有道路红线内利用标线形成弯曲的道路，同时抬升华乐路（建设三马路 - 建设六马路）沿线交叉口路面，实现机动车减速和片区交通宁静化。

Mark the road within the red lines to form a curved road. At the same time, uplift the intersections along Huale Road (Jianshesan Road–Jiansheliu Road) to achieve deceleration and reducing noise.

华乐路（建设四马路－建设五马路）取消垂直道路停车，设置为路内平行道路划线停车，释放人行空间。

Canceled the vertical roadside parking in Huale Road (Jianshesi Road–Jinshewu Road) and set parallel roadside parking to release more pedestrian space.

华乐路现状自行车交通量较大，建设六马路－淘金路段设置单侧独立自行车道，其余路段自行车与机动车混行，设置减速标志。

Currently, the bicycle traffic flow in Huale Road is large. Jiansheliu Road to Taojin Road section can be set independent bicycle lane, and other road section are set deceleration signs as the bicycles are mixed with motor vehicles.

华乐路（淘金路－青龙坊路）拓宽人行道，设置护柱，防止机动车停车占用行人空间。

Widen the sidewalk in Huale Road (Taojin Road– Qinglongfang Road), and set bollards to prevent motor vehicles form entering pedestrian space.

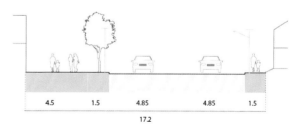

改造前
Before renovation

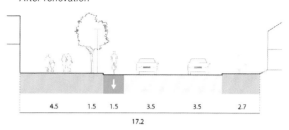

改造后
After renovation

华乐路（建设六马路－淘金路）断面
Huale Road (Jiansheliu Road-Taojin Road)senction

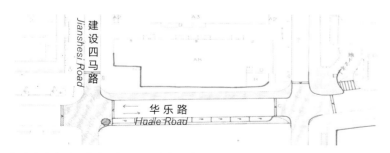

华乐路（建设四马路－建设五马路）方案
Huale Road (Jianshesi Road to Jianshewu Road) scheme

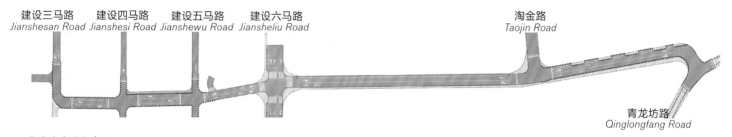

华乐路平面方案 The plane design of Huale Road

3.2.10 淘金东路
Taojindong Road

现状 Current Situation

①淘金东路为现状淘金道路网络中放射性的道路，双车道，沿街为菜市场及小商铺，人流较为密集，但很多路段人行道有效宽度不足 2 米。
②据调查，淘金东路晚高峰每小时标准车当量约为 500。

① Taojindong Road is the two-way radialized road inside the Taojin Road network. There are food markets and stores along the road with dense people, but the width of many sidewalk sections is less than 2m.
② According to the survey, the motor vehicles flow in evening peak hours in Taojindong is 500pcu/h.

现状
Current situation

设计方案 Design Plan

抬升路面，采用人行道铺装，设置护柱分隔。车行道宽为 5.5 米，双向通行。

Uplift and paved sidewalk, set bollard as separation. The roadway width is 5.5 meters, two-way traffic.

改善案例
Reference

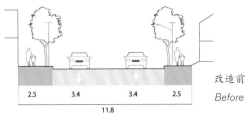

改造前
Before renovation

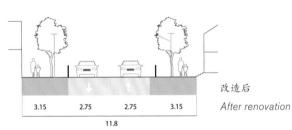

改造后
After renovation

淘金东路断面
The road section in Taojindong

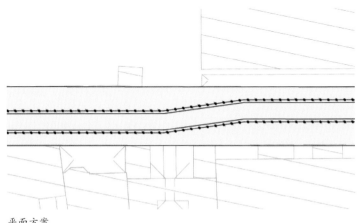

平面方案
Plane Scheme

3.2.11 淘金路
Taojin Road

现状 Current Situation

①淘金路现状为双向四车道，为淘金片区主要对外通道。淘金路－淘金北路北接恒福路，南承华乐路。淘金北路现状为双道，且淘金北路－恒福路交叉口，淘金北路往恒福路左转方向绿灯时间较短，通行能力受到限制。
②据调查，淘金路晚高峰每小时标准车当量约为1100。
③淘金路沿线均为商业和金融办公区域，人流量较大，现状人行道有效宽度均大于2米，但仍较为拥挤。
④淘金路中间以护栏分隔，人行过街绕行距离较长。

① The current Taojin Road is a two-way traffic lane. It is the main passage of Toajin area to other area. The north of Taojin-Taojinbei Road connects to Hengfu Road, and the south connects to Huale Road. The current Taojinbei Road is a two-way traffic lane, and the time of green light of the intersection in Taojinbei Road to Hengfu Road and the left turn of Taojinbei to Hengfu Road, is short, thus the passing capacity is limited.
② According to the survey, the traffic flow in evening peak in Taojin Road is about 1100pcu/h.
③ The road along Taojin Road is mainly business and financial office area, and the traffic is large. The effective width of the pavement is more than 2m, but it is still crowded.
④ There are bollards in the middle of Taojin Road, and the pedestrian detour distance is long.

在现状两个斑马线的基础上，再增设三个过街斑马线，减少绕行。

Basing on the current zebra line, set three zebra crossings in the pedestrian crossing to reduce detour.

Chapter Three

允许行人随意穿行的中央分隔带

Allow pedestrians to randomly across the central separation zone.

设计方案 Design Plan

调整为双向双车道（公交站点处为三车道），中间设置2米中央分隔带，取消护栏，释放人行空间。增加多处人行过街斑马线，减少绕行。

Adjust the road to two-lane road (three-lane road at bus stations)and set central separation belt in 2 meters long. Cancel the rail to release space for pedestrians. Add zebra crossing at various places to reduce detours.

改造前
Before renovation

改造后：普通路段
After construction: common section

改造后：公交站点处
After construction: bus stop

淘金路断面
Taojin Road Section

淘金路断面
Taojin Road Section

淘金路平面方案 *The plane scheme in Taojin Road*

206

3.2.12 青龙坊路
Qinglongfang Road

现状 Current Situation

①青龙坊路现状双车道，其中东段为双向通行，西段为单向通行，并设有路内停车。现状主要路段人行道有效宽度均不足2米。
②据调查，青龙坊路晚高峰每小时标准车当量约为250。道路空间未得到有效利用。

① Qinglongfang Road currently has two lanes, two directions on the east section and one way on the west section. There is no on-street parking. The effective width of the sidewalks on the majority of the road are currently less than 2 meters.
② According to the survey, the motor vehicles' flow in evening peak is 250pcu/h. The road space has not been effectively utilized.

现状
Current situation

改善案例
Reference

设计方案 Design Plan

抬升路面，采用人行道铺装，设置护柱分隔。车行道4米，单向通行。

Uplift and paved for sidewalk, and set bollard as separation. Set the traffic lane in 4 meters one-way lane.

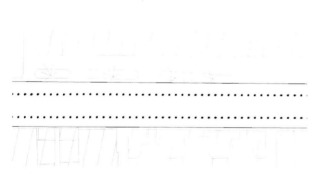

平面方案 *Plane Scheme*
人行道铺装，设置护柱分隔
Paved for sidewalk and set bollard as separation

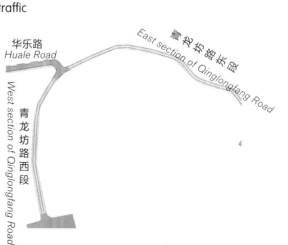

青龙坊路平面方案 *The plane scheme in Qinglongfang Road*

Chapter Three

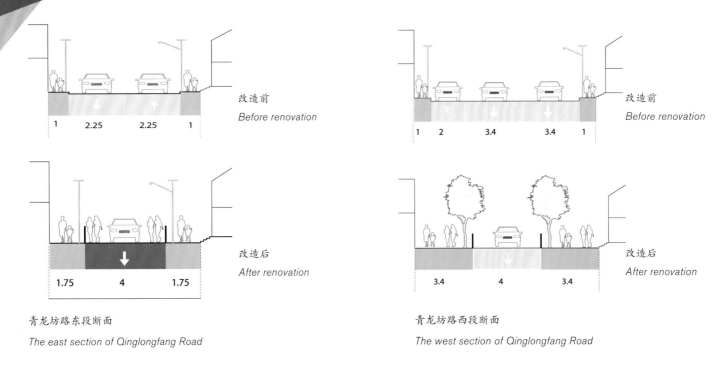

改造前
Before renovation

改造后
After renovation

青龙坊路东段断面
The east section of Qinglongfang Road

青龙坊路西段断面
The west section of Qinglongfang Road

3.2.13 建设六马路
Jiansheliu Road

现状 Current Situation

建设六马路现状为双向四车道，人行道有效宽度均大于 2 米。但由于建设六马路两侧有较多商业，上下班高峰期人流量较大。

The current Jiansheliu Road is a two-way four traffic lane, and the effective width of the sidewalk is wider than 2 meters. However, as both sides of Jiansheliu Road have relatively more business area, the population flow in peak hours is large.

调查方法：将建设六马路（环市东路至华乐路段）分成四段，每段各安排一名调查员进行 20 分钟的观测，记录下视野范围内所有行人过街的大致行走轨迹并标示在地图上。白色斑马线带代表现状的斑马线，黄色线条代表行人过街轨迹，线宽代表过街人数。

Survey method: Divide Jiansheliu Road (Huangshidong Road to Huale Road) into four sections, and each section has an investigator for 20 minutes observation to record the pedestrian crossing trajectory within the scope and mark it on the map. The white zebras represents the existing crosswalk, and the yellows line demonstrate pedestrian crossing trajectory. The width of the lines represents the number of pedestrians.

实践案例
Practice

调查结果如下：

The results are as follow:

①横过建设六马路（环市东路至华乐路段）的行人过街设施仅有两处：宜安小区出入口、华乐路交叉口；
②环市东路与建设六马路交叉口缺少横过建设六马路的设施。现状路口涂画有三角形导流带，导流带成为了行人过街的中转点。
③宜安小区门前的斑马线及斑马线周边行人过街密度很高。
④宜安小区至华乐路之间，虽然没有过街设施但行人过街行为频繁，尤其是麦当劳餐厅以及六街咖啡路段是行人聚集的过街路线。
⑤华乐路口行人过街流量较大。
总结：仅仅两处行人过街设施，无法满足建设六马路（环市东路至华乐路段）沿线大量的过街需求。

① There are only two pedestrian crossing facilities across Jiansheliu Road (Huanshidong Road to Huale Road). They are Yian neighborhood and intersection of Huale Road.
② The intersection in Huanshidong Road and Jiansheliu Road is lack of road crossing facilities. The current junction is painted with triangle diversion belts which become the transit point of pedestrian crossing.
③ The intensity of zebra line and the pedestrians flow in front of the entrance of Yian neighborhood is high.
④ Although there is no road crossing facility in Yian neighborhood between Huale Road,the pedestrian frequently cross the street, especially the road crossing at Macdonalds and the café road section in Jiansheliu Road.
⑤ The pedestrian crossing flow is large in Huale Road.
Conclusion: Only two pedestrian crossing facilities are not enough and cannot meet the demand of large pedestrian crossing along Jiansheliu Road (Huanshidong Road to Huale Road).

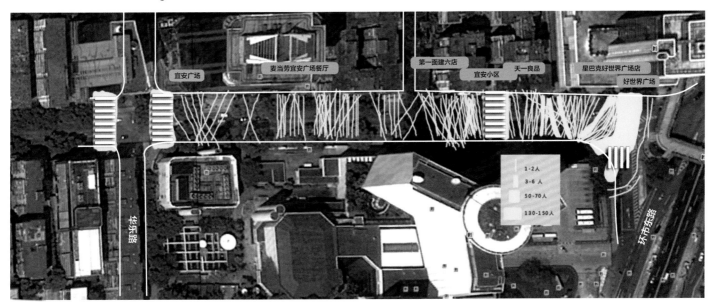

行人过街轨迹图 Pdeestrian crossing trajectory

Chapter Three

设计方案 Design Plan

①建设六马路（环市东路－华乐路）调整为双车道，中间设置2米中央分隔带，允许行人随意穿行，其余道路空间释放给行人。建设六马路（华乐路－东风路）调整为双向三车道，其中北往南方向为两车道，南往北方向为单车道，中间设置2米中央分隔带，允许行人随意穿行，其余道路空间释放给行人。
②建设六马路－环市路交叉口进行渠化，增加行人过街安全岛。

① Adjust Jiansheliu roadway (Huanshidong Road-Huale Road) to two-way traffic lane, and set 2 meters separation belt in the middle. Allow pedestrians to randomly cross and release more space for pedestrians. Adjust Jiansheliu Road (Huale Road to Dongfeng Road) to two-way three lanes, among which the two-way lane drives from north to south direction, while the one-way traffic lane drives from south to north direction. Set 2 meters central separation belt, and allow pedestrians to randomly across the road, the rest of the space released for pedestrians.
② Channelize the intersection in Jiansheliu Road to Huanshidong Road. Add more pedestrian safety islands.

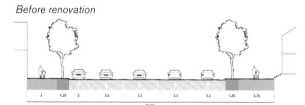

改造前
Before renovation

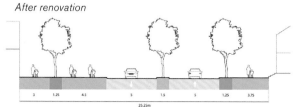

改造后
After renovation

建设六马路（环市东路－华乐路）断面
The intersection in Jiansheliu Road (Huanshidong to Huale Road)

改造后
After renovation

建设六马路（华乐路－东风路）断面
The road section in Jiansheliu Road (Huale Road to Dongfeng Road)

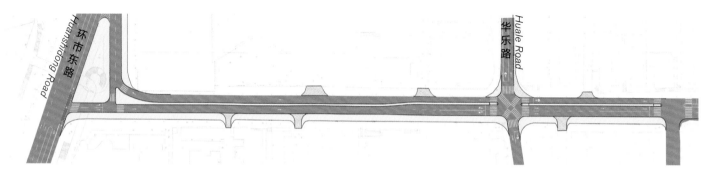

建设六马路平面方案 The plane shecme in Jiansheliu Road

实践案例
Practice

现状

Current situation

现状

Current situation

改善案例

Reference

建议

Proposal

211

Chapter Three

3.2.14 环市东路
Huanshidong Road

现状 Current Situation

①环市东路（建设大马路 – 建设六马路）沿线较多后退区停车。
②据调查，环市东路晚高峰时期自行车流量较大。

① There are many back area parking places along Huanshidong Road (Jianshe Avenue to Jiansheliu Road).
② According to survey, the bicycle flow in evening peak is large in Huanshidong Road.

设计方案 Design Plan

①沿线出入口路面抬升，并设置护柱，防止后退区停车。
②在辅道设置自行车道，划线分隔。同时设置少量停车位，以满足沿线单位临时停车需求。

① Uplift the entrance along the road; set bollard; prevention of back area parking.
② Pave bicycle lane on pavement, and divide by lines. At the same time, set small amount of parking area to satisfy the parking needs of the apartment along the road.

现状
Current situation

建议
Proposal

实践案例
Practice

改造前
Before renovation

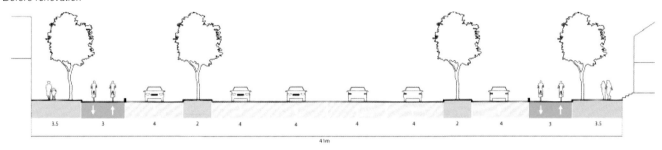

改造前
Before renovation

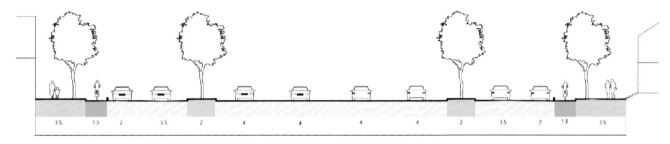

环市东路断面
The road section in Huanshidong Road

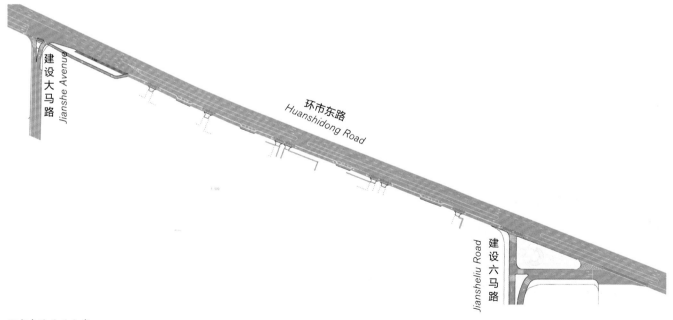

环市东路平面方案
The plane scheme in Huanshidong Road

Chapter Three

老挝万象绿道规划设计
Greenway Planning and Design in Vientiane, Laos

目标
Objectives

此次研究及初步设计的主要目的是为即将实施的亚行贷款项目——万象非机动车交通（NMT）改善项目打下基础。

The main objective of this study and preliminary design is to provide a basis for the implementation of improvements for non-motorized transport (NMT) in Vientiane, preferably as part of a coming Asian Development Bank (ADB) loan.

次要目标是为亚行除万象外的另外两个城市，棉兰和马尼拉的非机动车交通（NMT）改善项目提供借鉴。

A secondary objective is to provide input to the ADB study on NMT in three case study cities in Asia; the other two being Medan and Manila.

 当前问题
Current Problems

数个问题造成了万象步行和骑行条件不佳。

Several issues create poor conditions for walking and cycling in Vientiane.

2.1 缺失及不连续的人行道
Missing and Discontinuous Walkways

现状只有少数街道包括 Samsenthai、Setthathilath 和 Quai Fa Ngum，将良好的人行道作为道路建设的一部分。其他许多街道或部分或完全缺少人行道，或者人行道太窄以至于人们只能选择走在机动车道上。此外，通向建筑物和停车场的机动车出入口也频繁切断现有人行道，造成行人的不便，错误的体现了机动交通路权的优先。机动车出入口也为违规停车，阻碍行人通行。

A number of streets, including Rue Samsenthai, Rue Setthathilath and Quai Fa Ngum have raised sidewalks as part of road reconstruction projects. Many other streets lack sidewalks partly or completely or have sidewalks so narrow that walking on the street is the only option. Moreover, driveways, small streets for building access which lead into courtyards and parking lots, force pedestrians to step down and up the sidewalk, causing inconvenience and wrongly indicating motorized traffic priority. Driveways also provide opportunities for illegal parking, blocking pedestrians' thoroughfare.

实践案例
Practice

狭窄的人行道迫使行人走在街道上。

Narrow sidewalks force pedestrians to walk on the street.

缺少人行道使得街道的步行条件不佳。

Streets without sidewalks result in poor walking conditions.

机动车出入口切断人行道，人行道不连续，并让机动车更容易驶进人行道。应提升机动车出入口，并设置护柱阻止机动车驶入。

Driveways interrupt the continuity of sidewalk and provide easy access for motorized vehicles to entersidewalks. Driveways need to be raised and bollards are needed to prevent vehicle encroachment.

2.2 障碍物
Obstacles

大部分万象人行道上的障碍物对行人而言不仅造成出行不便，还会危及安全。这些障碍物包括公共设备、交通标识、破损的路面、突出物、破洞、广告牌、街边小贩和垃圾堆等。

Obstacles on most of Vientiane's sidewalks pose a danger and inconvenience to pedestrians. These include utilities, traffic signs, broken pavements, protrusions, holes, advertisement boards, vendors and rubbish.

2.3 人行道上的停车
Parking on Sidewalks

由于执法力度不严，人行道上乱停车现象非常普遍。

Due to poor enforcement, parking on sidewalks is excessive.

2.4 交叉口
Intersections

万象交叉口的设计中存在诸多问题。因为设置了不必要的大缘石半径，所以交叉口半径变得过大，导致行人过街时面对高速转弯的汽车只能一直等待过马路时机。其它问题则表现在缺乏人行横道，以及交叉口处停车阻塞人行道等方面。改善这些问题将为所有道路使用者带来好处。

The designs of Vientiane's intersections have several issues. Due to unnecessarily large curb radii intersections are too wide, resulting in high car turning speeds and pedestrians crossings which are set back. Other issues are a lack of pedestrian crossings and parking at the intersection, blocking pedestrian thoroughfare. Improvements would benefit all road users.

2.5 人行横道
Pedestrian Crossings

缺少路中过街通道或路中过街通道设计较差。

Mid-block pedestrian crossings are often lacking or poorly designed.

2.6 缺乏遮盖设施
Lack of Shade

许多道路都缺乏遮盖设施来应对烈日和雨水，使得行人丧失步行的积极性。

Protection from sun and rain is lacking on many roads, discouraging people from walking.

建议书
Proposals

3.1 人行道
Sidewalks

3.1.1. 人行道及路面
Sidewalks and Pavement

一些街道缺乏人行道，许多道路的人行道上现有的铺装已经损坏。右上图显示人行道的现状，包括人行道缺失和人行道铺装损坏程度。右下图显示了建议改善的人行道铺装的位置及具体的工程量。

Several streets lack sidewalks and the pavement on many existing sidewalks is damaged. The Missing sidewalks and pavement damage Proposed reconstruction of sidewalks graph below shows the state of existing sidewalks. The graph on the following page shows where sidewalks and pavement are proposed to be (re) built.

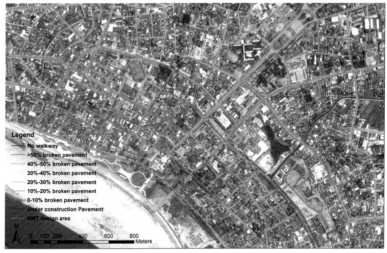
缺失的人行道和人行道铺装破损程度
Missing sidewalks and pavement damage

人行道的改造建议
Proposed reconstruction of sidewalks

Chapter Three

3.1.2. 隔离
Separation

建议在人行道边缘设置护柱，以防止车辆的驶入。拟设置三种类型的护柱，护柱规格如下：

Bollards are proposed on the edge of sidewalks to prevent vehicle encroachment. The bollard specifications are shown below. Three types are proposed:

①设置在交叉口和路中过街的护柱；
②设置在道路路段的护柱，此类护柱可以选择较粗护柱，可兼作公共座椅；
③设置在道路路段的护柱，中间用铁链连接，这些护柱可以兼做公共座椅并能阻止摩托车驶上人行道。

① Bollards at intersections and mid-block crossing points
② Bollards at mid-block section; these can double as public seating
③ Bollards at mid-block section with chain; these can double as public seating and keep motorcycles off sidewalks.

阿姆斯特丹（左图）和里昂（右图）的护柱
Bollards in Amsterdam (left) and Lyon (right)

3.1.3. 障碍物
Obstructions

万象的人行道经常被各种固定或可移动的障碍物所占用。下图显示了建议移除或者重新安置的阻碍人行道的交通标识和灯杆，其中一些标识高度不足，只到成年人的头部高度，对路人的安全有一定的威胁。

Vientiane's sidewalks are often obstructed by various fixed and movable objects.The following graph shows proposed removal and relocation of signs and light poles which obstruct the walkway, with signs often at a hazardous head-height.

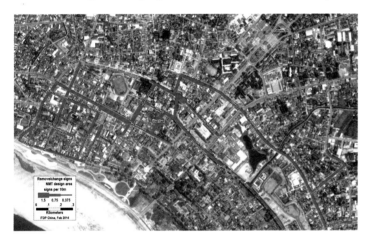 

建议移除和重新安置的交通标识（左图）和灯杆（右图）
Proposed removal and relocation of traffic signs (left) and light poles (right)

3.1.4. 机动车出入口和小支路路口
Driveways and side streets

人行道经常在机动车出入口和小支路路口处被切断，行人被迫在切断处上下人行道。在车流量非常低的机动车出入口和小支路，保持人行道的连续是首选，这使得步行更加便利并能降低车辆通过速度。建议提升机动车出入口和小支路至人行道平面。

Sidewalks are often cut-off at driveways and side-street and pedestrians need to step down and up the curb. At driveways and smaller streets, where traffic volumes are very low, continuous sidewalks are preferred. This makes walking more convenient and slows down vehicle traffic.Propose raised sidewalks at driveways and side streets are provided.

3.2 交叉口和过街通道
Intersections and Crossings

为行人和骑行者创造连续的道路网，保持交叉口处人行道和自行车道的连续是必需的。研究范围内所有交叉口双车道以上和机动车流量大的街道都上都应设置行人安全岛。设计转弯小半径，扩宽行人空间，并在适当地点外延人行道以确保机动车低速转弯通过交叉口，从而让行人和自行车更加安全地过街。一些更小的交叉口建议抬升，以降低行人密集区域的车速。在路段设置带有安全岛的路中过街通道，与非机动交通网络相匹配。

To create continuous networks for pedestrians and cyclists intersection treatments are needed. The designs of all intersections within the scope area include pedestrian refuge islands for crossings of more than two lanes of traffic and on streets with high traffic flows. Curb radii are minimized, pedestrian space widened and, where appropriate, bulb-outs included to ensure slow turning of motorized traffic making pedestrian and bike crossings safer. Several smaller intersections are proposed to be raised to slow down traffic in popular pedestrian areas and sidewalks are proposed to be raised through small intersections and driveways. Mid-block crossings with refuge islands are proposed to ease crossing between intersections and align with NMT path entrances.

Quai Fa Ngum - Mantathourath 交叉口的现状和改善建议：连续的人行道使得步行穿过交叉口变得安全而便利。外延人行道和设置护柱为行人创造了更多空间，缩短了过街距离，同时防止车辆侵占人行道。

Quai Fa Ngum - Rue Mantathourath. The continued sidewalk makes walking through the intersection safe and convenient. The bulb-out and bollards create more space for pedestrians, reduce the crossing distance and prevent car encroachment.

实践案例
Practice

3.3 道路设计和宁静交通
Road Layouts and Traffic Calming

对研究区域道路逐条进行宁静交通设施设计，这些设计包括：
①减速路拱和减速台
②减少路缘石转弯半径
③路面的曲折和缩窄
④减少机动车道
⑤抬升交叉口
⑥交通组织改善

Many traffic calming measures are incorporated into the street-by-street proposals. These include:
① Speed humps and tables
② Narrowing of turning radii
③ Chicanes and narrowing of roadways
④ Lane reductions
⑤ Raised intersections
⑥ Traffic circulation changes.

荷兰乌特勒支的减速台/减速路拱
Speed table/hump in Utrecht, The Netherlands

3.4 街道家具和照明设备及景观美化
Furniture and Light and Landscaping

人行道上靠近道路缘石的区域被称为"街道家具区",灯杆、绿化、交通标识、消防栓、座椅、自行车停放处、果皮箱等通常被安置在这一区域。然而在万象的研究区域,人行道通常没有足够的宽度来设立一个街道家具区,因此道路的景观美化、座椅设置和其他用途设施非常有限。然而也有例外情况,部分道路拓宽后可设置明确且足够宽的家具区以适应各种各样的需要。Lane Xang 大道就设置了相当充裕的"街道家具区",这是少数街道才有的。然而区域内的一些人行道在拓宽后也能设置"街道家具区"(在宽度足够的区域),这将有助于种植树木、树立灯杆、街道标识、设置座椅、自行车停放架和垃圾箱等设施。

The area of the sidewalk closest to the curb, where light poles, plantings, signs, fire hydrants, seating, bike parking, waste receptacles, etc., are typically located, is referred to as the "furnishing zone." In the study area in Vientiane, however, sidewalks are generally not wide enough to accommodate a furnishing zone, so opportunities for landscaping, seating and other uses within this zone are very limited. There are exceptions, however, where the furnishing area is or will be defined and wide enough to accommodate various usages. Ave. Lane Xang has a fairly well defined "furnishing areas", but few other street segments do. Several walkways in the study area will be widened, however, and this widening will create an opportunity to better define a "furnishing area" (where the width allows one), which in turn can guide the placement of trees, light poles, street signs, seating, bike parking racks, rubbish receptacles, and so on.

布达佩斯一条街道上的家具设置。

Street furniture on a street in Budapest.

实践案例
Practice

3.5 公共座椅
Public Seating

在某些街道，一般在拓宽人行道和移除障碍物而得以划分出家具区之后，可设置公共座椅，为行人步行和休闲营造更舒适的环境。

On certain streets, generally where a furnishing zone is available after walkway widening and obstacle removal, public seating is proposed to provide a more conducive environment for walking and leisure.

3.6 树木
Trees

为了改善步行条件，建议在一些街道上种植树木为人行道提供遮阳，主要集中在暴露在太阳底下时间最长东西走向而又几乎没有建筑物遮蔽的街道上。尽管拓宽后人行道的宽度理论上能够栽种更多树木，但种植树木的位置还是受到道路和人行道宽度的严重制约。一般而言，除了 Hengboun 路拟种一排树木之外，要避免在商铺林立的街道前面植树。在步行网络的部分路段也拟种树木，特别是一些关键的连接道路，这些路段都是东西走向，且道路临街面活动不活跃。在项目预算中，栽植树木的维护成本在第一年将被并入树种的购置费用中。树木在最初时候需要每天浇水，但即便如此，预计最多会有 30% 的树木不能成活，而如果维护差一点，成活率可能只有 50%。

To improve walking conditions, trees are proposed to be placed on several streets, focusing on east-west streets which have the greatest exposure to the sun throughout the day, with the least shade from buildings. Placement of trees is also heavily constrained by the road and walkway width, though the proposed walkway widening makes it possible to plant more trees. In general tree planting was avoided in front of streets with dense rows of shops, except for Rue Hengboun where a row of trees is proposed. Trees were also proposed in parts of the pedestrian network with east-west orientation and alignment along walls or other inactive streetfronts, especially in key connecting street sections. The maintenance cost of establishing the trees over the first year has been incorporated into the tree cost in the project budget. Trees will initially require daily watering and even then it is expected that up to 30% will not survive to maturity. With poorer maintenance, the survival rate may be only 50%.

作者简介
About the Authors

1. 刘少坤 ShaoKun Liu

交通与发展政策研究所中国区（ITDP-China）副主管，致力于推动中国城市可持续交通发展，为交通与发展政策研究所中国区慢行交通、绿道、停车管理项目负责人，同时也是快速公交（BRT）系统、公共自行车项目的咨询专家。自2009年起，参加了中国数十个城市的可持续交通项目，也作为世界银行、亚洲发展银行交通专家在中国、马来西亚、菲律宾、老挝、印度尼西亚、蒙古等国参加了数十个可持续交通项目的工作，参与的绿道规划建设项目遍布国内外数十个城市，对中国及东南亚地区的绿道面临的挑战及发展趋势有深入的研究。

Shaokun Liu is the Vice Country Director of ITDP-China (Institute for Transportation & Development Policy), he has been promoting sustainable and equitable transportation in China, and was ITDP-China NMT, Greenway, Parking and TDM project team leader. Since 2009 He has been working on urban and sustainable transportation projects in several China cities. He has been worked as a consultant to Asian Development Bank and World Bank on many sustainable transportation projects in China, Malaysia, Philippines, Laos, Indonesia, Mongolia, etc., and has working on greenway planning & design in several cities in worldwide. He knows the challenges and opportunities for Greenway developing in China and South East Asia cities very well.

2. 黎淑翎 Shuling Li

国家注册城市规划师，英国曼彻斯特大学规划硕士，（英国）皇家城镇规划协会（RTPI）执业会员（Licentiate Member），是交通与发展政策研究所中国区（ITDP-China）城市发展项目咨询专家，也是亚洲开发银行在册的咨询专家。自2011年起，先后参与了广州、兰州、宜昌、天津、老挝万象、马来西亚新山等城市的可持续交通项目，并与国内外机构和专家共同编写、翻译和推广《珠三角城市发展最佳实践》、《公交导向发展评价标准》等研究报告和规划导则，专注于BRT走廊沿线的公交导向发展的政策研究与规划设计，尤其是在慢行系统、绿道及停车与土地利用规划、城市设计的相互协调方面，有丰富的研究及设计经验。

With the degree of Master of Planning in University of Manchester (UK), Shuling Li is a National Certified Urban Planner (China), and a Licentiate Member in Royal Town Planning Institute (RTPI, UK). She is a consultant for the Urban Development Program in ITDP China office and Asian Development Bank. Since 2011, Shuling has taken a major role in the sustainable transportation and planning projects in several cities, including Guangzhou, Lanzhou, Yichang, Tianjin, Vientiane (Laos) and Johor Bahru (Malaysia). Meanwhile, she has cooperated with national and international organizations and experts in writing, translating and promoting research reports and planning guidelines, including Best Practices in Urban Development in the Pearl River Delta and TOD Standard. Shuling focuses on policy research, planning and design for transit-oriented development along BRT corridors, and has extensive research and practical experience in the integration with non-motorized transport, greenway, parking, land use planning and urban design.

3.Bram van Ooijen(欧阳白)

荷兰人，荷兰格罗宁根大学城市规划及地理信息系统学士，荷兰屯特大学土木工程及工程管理硕士，是交通与发展政策研究所中国区(ITDP-China)慢行交通及停车管理的规划设计及政策的咨询专家。自2010年起，欧阳白作为技术专家，积极与中国国家、省、市、区各级政府，世界银行、亚洲开发银行、联合国开发计划署等国际机构及国内设计单位合作，在中国可持续交通项目的规划与设计，尤其是在停车管理和绿道方面，有着深入的研究和丰富设计成果。他在国内外工作的城市包括北京、广州、天津、宜昌、马尼拉、雅加达、万象和新山等。

Bram van Ooijen works on policy, planning and design of non-motorized transport infrastructure and parking systems at the Institute of Transportation & Development Policy (ITDP). Based in Guangzhou, China, since 2010 he works with national, provincial, municipal and district-level politicians and government bureaus, as well as planning and design institutes and multi-lateral development banks. He is involved in planning, design and implementation of greenways, bike networks, street design and pedestrian and bike access to transit (BRT) systems. He worked in major Chinese cities such as Guangzhou, Beijing, Tianjin, Harbin, Lanzhou, Yichang and Foshan and Asian capitols including Manila, Jakarta, Vientiane and Johor Bahru. Bram van Ooijen is also involved in advocacy and training of stakeholders in the topics of non-motorized transport and parking as a consultant, presenter, trainer and author of reports and articles for governments, multi-lateral development banks (a.o. World Bank, Asian Development Bank, United Nations Development Program) and NGOs. He graduated from Twente University with a Master of Science in civil engineering & management, specialized in transportation, and a Bachelor of Science in Civil Engineering.

4. 李珊珊 Shangshan Li

交通与发展政策研究所中国区（ITDP-China）副主管，致力于推动中国及区域范围内可持续和平等的城市交通，是交通与发展政策研究所中国区快速公交（BRT）项目、公共自行车系统项目负责人，同时也是慢行交通、城市设计项目的咨询专家。自2009年起，参与了数十个城市的可持续交通项目，同时作为世界银行、亚洲发展银行的交通咨询专家参与了其在中国及东南亚地区的多个交通项目，包括中国的广州、兰州、宜昌、天津等，马来西亚的吉隆坡及新山，菲律宾马尼拉，老挝万象，印度尼西亚雅加达等城市。并多次参加亚洲发展银行的城市交通论坛及其他国际交通论坛，对城市快速公交及慢行交通发展有深入的研究及探索。

Shanshan Li is the Vice Country Director of ITDP-China (Institute for Transportation & Development Policy). She joined ITDP since 2009 after graduated from Kunming University of Science and Technology. She has been working on urban and sustainable transportation projects, and mostly on BRT and NMT. She has been worked as a consultant to Asian Development Bank on many sustainable transportation projects in China and Asian cities. She has been working on Yichang BRT and NMT planning, design, implementation and operation project (Yichang BRT system opened at July, 2015), and she has been working on sustainable transportation projects in Lanzhou, Guangzhou, Tianjin, Vientiane, Kuala Lumpur, Jakarta and Johor Bahru.

5. 李薇 Wei Li

华南理工大学交通工程专业毕业。交通与发展政策研究所中国区（ITDP-China）快速公交（BRT）及慢行交通咨询专家、世界银行及亚洲开发银行咨询顾问、高级工程师。自2005年，起致力于在世界范围内，尤其是东亚及东南亚推广绿色可持续交通出行模式。参与并完成了广州、宜昌、天津、蒙古乌兰巴托、马来西亚吉隆坡和新山、印尼雅加达、老挝万象、菲律宾马尼拉等多个国内外可持续交通改善项目。

Wei Li is a traffic and transportation engineer. She has worked mostly on Bus Rapid Transit (BRT) and non-motorized transport (NMT) since graduating from the South China University of Technology since 2005. As well as working for ITDP, she has worked as a consultant to both Asian Development Bank and World Bank-funded urban transport projects in China and Asian cities, and she was part of ITDP's team working on the sustainable transport project in Guangzhou, Yichang, Tianjin, Ulaanbaatar, Kuala Lumpur & Johor Bahru, Jakarta and Vientiane.

图书在版编目（CIP）数据

城市绿道系统优化设计 / 交通与发展政策研究所（中国办公室）编著 . -- 南京：江苏凤凰科学技术出版社，2016.9

ISBN 978-7-5537-6923-3

Ⅰ . ①城… Ⅱ . ①交… Ⅲ . ①城市道路－道路绿化－绿化规划 Ⅳ . ① TU985.18

中国版本图书馆CIP数据核字（2016）第 172787 号

城市绿道系统优化设计

编　　　著	交通与发展政策研究所（中国办公室） Institute for Transportation & Development Policy (ITDP-China)
项 目 策 划	凤凰空间 / 罗瑞萍　官振平
责 任 编 辑	刘屹立
特 约 编 辑	官振平
出 版 发 行	凤凰出版传媒股份有限公司 江苏凤凰科学技术出版社
出版社地址	南京市湖南路1号A楼，邮编：210009
出版社网址	http://www.pspress.cn
总　经　销	天津凤凰空间文化传媒有限公司
总经销网址	http://www.ifengspace.cn
经　　　销	全国新华书店
印　　　刷	上海利丰雅高印刷有限公司
开　　　本	787 mm×1092 mm　1 / 12
印　　　张	19
字　　　数	182 400
版　　　次	2016年9月第1版
印　　　次	2016年9月第1次印刷
标 准 书 号	ISBN 978-7-5537-6923-3
定　　　价	288.00元（USD 55.00）（精）

图书如有印装质量问题，可随时向销售部调换（电话：022-87893668）。